EDUCATION IN A COMPETITIVE AND GLOBALIZING WORLD

EMPOWERING SCIENTIFIC LITERACY THROUGH DIGITAL LITERACY AND MULTILITERACIES

EDUCATION IN A COMPETITIVE AND GLOBALIZING WORLD

Additional books in this series can be found on Nova's website under the Series tab.

Additional E-books in this series can be found on Nova's website under the E-book tab.

EDUCATION IN A COMPETITIVE AND GLOBALIZING WORLD

EMPOWERING SCIENTIFIC LITERACY THROUGH DIGITAL LITERACY AND MULTILITERACIES

WAN NG

Nova Science Publishers, Inc.
New York

For permission to use material from this book please contact us:
Telephone 631-231-7269; Fax 631-231-8175
Web Site: http://www.novapublishers.com

Additional color graphics may be available in the e-book version of this book.

LIBRARY OF CONGRESS CATALOGING-IN-PUBLICATION DATA

Empowering scientific literacy through digital literacy and multiliteracies / editor, Wan Ng.
p. cm.
Includes bibliographical references and index.
ISBN 978-1-62100-768-5 (hardcover)
1. Science--Study and teaching. 2. Technological literacy. 3. Science--Methodology. I. Ng, Wan, 1955-
Q181.E625 2011
507.1--dc23

2011038506

Published by Nova Science Publishers, Inc. † New York

CONTENTS

PREFACE

We live in a society largely driven by science and technology. As more scientific and technological issues dominate public debates at national and international levels, it is important to ensure that our students become global citizens who are scientifically literate. However, many students have poor attitudes and low engagement levels toward the learning of science. This book puts forward an argument that we should capitalise on the affordances that digital technologies offer in enabling better science learning, the general technological interest and knowledge of young people and the motivating influence of technology for learning, to foster the development of scientific literacy in students.

Chapter 1 - This book puts forward a case that educators should capitalise on (i) the affordances that digital technologies offer in enabling better science learning (ii) the general technological interest and knowledge of young people and (iii) the motivating influence of technology on learning, to foster the development of scientific literacy in students. The book argues that ensuring that students are digitally literate would empower them to embrace digital resources to learn science better and to use effectively the resources to create artefacts that demonstrate their understanding of concepts learned and/or for the synthesis of new ideas. An important line of argument for digital literacy is that its proficiency would alleviate cognitive load when students learn science in technologically-enriched environments so that their minds could focus on the content or task at hand rather than having to split their attention with the technological aspects of the learning. An implication of digital literacy in science learning is that science teachers need to be similarly digitally literate in integrating digital technologies in their teaching. The book also has a focus on the multiliteracies perspective of digital literacy, that is, the multimodal means of learning and communicating in the development of scientific literacy.

Chapter 2 - This chapter defines scientific literacy and introduces the concept and its development at two levels, a specialist level and a generalist level. It will discuss the importance of being scientifically literate and the skills and knowledge associated with it. Based on the literature and other reports around the world, the chapter will present the theoretical dimensions and frameworks for scientific literacy proposed by researchers and establish a position for the purpose of this book. It will look beyond the classroom development of scientific literacy and discuss informal learning that takes place outside the classroom, particularly learning that is facilitated by digital technologies.

Chapter 3 - Similar to the argument for being scientifically literate is the argument that students and the general public should be digitally literate. This is because digital technology tools are advancing and proliferating the marketplace at an increasing pace. Their presence is everywhere in our daily lives and they are becoming more central to the individual's learning and social wellbeing. They are also central to the economic development and advancement of businesses, corporations and nations. E-Skills UK (2009) reported on the increasing number of graduates directly employed in 'digital industries', with an estimated 77% of UK jobs involving some level of competency in ICT. The requirement of workers to update skills as technologies get upgraded is a common feature of workplaces in the 21st century. The importance of digital literacy in terms of the economy and employment rates has been stressed by Wynne and Cooper.

Chapter 4 - The influence of digital technologies on meaning making (the construction of understanding) in the last two decades has seen the concept of 'literacy' slowly evolving to the literacy plurality of 'multiliteracies' and the definition of language 'morphing into metalanguage' (Adwell et al., 2007). The term 'multiliteracies' was first proposed by the New London Group (1996) to highlight two arguments that are influenced by globalisation and technology: (i) the significance of cultural and linguistic diversity and (ii) the multiplicity of communication channels. The former argument asserts that as today's society becomes more linguistically and culturally diverse, meaning making differs according to cultural, social and professional contexts while the latter suggests that media and communication technologies enable meaning making to be increasingly multimodal and where the written-linguistic modes are integral of visual, audio, gestural and spatial patterns of meaning.

Chapter 5 - In the preceding chapters, I have discussed the three main components of this book: scientific literacy, digital literacy and multiliteracies. This chapter aims to tie these components together to develop a framework for the empowering of scientific literacy development through digital literacy and multiliteracies, the latter being the multimodality of representations enabled by the affordances of digital technologies.

Chapter 6 - In science education, the aims of using ICT are to support the learning of science concepts - through games and simulations; obtaining authentic science-related experiences such as the space projects at NASA website or virtual museum tours; enabling data collection through data logging equipment; communicating and discussing results and ideas through graphs, spreadsheets, email and possibly social networking communities, and synthesising and creating knowledge products such as a science digital story, blog or a wiki. As discussed in previous chapters, ICT-enabled affordances that support pedagogies in science education include a range of hardware and software. Examples of these are the interactive whiteboard, digital microscopes, simulations, interactive worksheets, electronic laboratory notebook, science websites, audio and video recording devices such as flipcams, multimedia editing tools, Web 2.0 technologies and data logging equipment and software. The ability to use these technologies for effective teaching and learning would require a reasonably level of digital literacy from both the teacher and the students. Teachers with an understanding that multimodality is a characteristic of scientific texts and that ICT enhances this aspect of the text (Knain, 2006) would be better prepared in their teaching to develop their students' scientific literacy. Knain's multimodality assertion is echoed by Prain and Waldrip.

Chapter 7 - Teaching is a complex activity and the process of preparing teachers to teach is similarly complex. The demands of a society that is very much influenced by the fast-changing nature of technology create additional challenges for teachers and teacher educators. In order to both reflect and engage with these changes, it is important for educators to be aware of the changing culture and expectations of the younger generation of students, who use ICT creatively to communicate and to seek and process information for both educational and recreational purposes. The challenges for teacher educators include staying abreast of new technologies and changes associated with them while simultaneously educating pre-service teachers on how to integrate technology effectively into their teaching methods. Similar challenges are faced by practicing teachers in the need to stay abreast of new educational technologies and integrating them appropriately into their teaching. ICT has the potential to offer rich learning environments that foster flexible learning and enable students to adopt multiple perspectives when learning about complex and abstract science phenomena. In order to harness this potential effectively, teachers need to be well prepared and possess a sufficient level of digital literacy to facilitate the learning. This chapter will discuss the issues associated with ICT integration in education and the implications for teacher adopting these technologies in the science classroom.

Chapter 8 - This book has put forward an argument that the development of scientific literacy could be enhanced if students and teachers are digitally literate and understand that multimodality – the representation of concepts in multiple modes is inherently associated with scientific literacy.

ACKNOWLEDGMENTS

I am grateful to Dr. Howard Nicholas for the discussions that we have had and for helping me clarify my thoughts throughout the writing of this book. Thank you for your support, encouragement and care.

To Charles Boshier, a talented Master of Education student and science teacher, thank you for your contribution to Chapter 6.

Thank you to Leni Suek for assisting with the editing aspect of this book.

To the many pre-service teachers and primary and secondary school students and their teachers whom I have worked with over the years, thank you for helping me grow in my understanding of science learning, with and without the use of digital technologies.

For Michael and Michelle, whom I love dearly
Thank you for being such good, caring and responsible children

Chapter 1

INTRODUCTION

PREAMBLE

This book puts forward a case that educators should capitalise on (i) the affordances that digital technologies offer in enabling better science learning (ii) the general technological interest and knowledge of young people and (iii) the motivating influence of technology on learning, to foster the development of scientific literacy in students. The book argues that ensuring that students are digitally literate would empower them to embrace digital resources to learn science better and to use effectively the resources to create artefacts that demonstrate their understanding of concepts learned and/or for the synthesis of new ideas. An important line of argument for digital literacy is that its proficiency would alleviate cognitive load when students learn science in technologically-enriched environments so that their minds could focus on the content or task at hand rather than having to split their attention with the technological aspects of the learning. An implication of digital literacy in science learning is that science teachers need to be similarly digitally literate in integrating digital technologies in their teaching. The book also has a focus on the multiliteracies perspective of digital literacy, that is, the multimodal means of learning and communicating in the development of scientific literacy.

WHY STUDY SCIENCE?

Apart from the reason that science is taught in schools to prepare for future scientists and engineers, I think that we should all enjoy science and think about the importance of science knowledge in a society that is largely driven by science and technology. Science permeates our culture and the effects of science have reshaped the world we live in. While not all that is science is good (for example, atomic bombs), it is undeniable that mankind is overwhelmingly better off now (for example, better health, better living conditions and lifestyle) than in pre-scientific times. There will be people who will assert that science and what flows from it is destructive, altering both the mental state and physical environment and producing stress and misery. Irrespective of whether science is seen as good or evil, it deserves attention because of its impact. Its impact can be seen in the increased crop yields, control of diseases, increased life spans, the diverse range of electrical and electronic gadgets that we use daily, transport infrastructure and vehicles (cars, ships and planes) and electronic communicative devices and platforms (phone, radio, television and via the Internet). These are aspects of our present environment that are so pervasive that we take them for granted. These developments occurred within the last 150-200 years and are considered very recent in terms of human history, which, depending on how you define 'human', have been shown by genomic evidence to go as far back as half a million years (Green, Krause, Briggs, Ronan and Simons, 2006) when Homo Sapiens first appeared, or about 100,000 years (Reader, 1988) when the first anatomically modern Homo Sapien existed. These developments are the result of science and the intertwined technologies.

Science and technology broadens possibilities. The results of these possibilities are present throughout our everyday lives. A good reason to know some science is to understand about these developments that are around us. Even during one lifetime the possibilities opened up by science for profound change in the way we live are enormous. The idea of change is often disturbing and even frightening to some people. Part of that anxiety results from having to face issues that appear daunting because they are incomprehensible. How the possibilities that science offers are used depends on societal decisions – decisions made by individuals or by groups, as family, as workplaces, as companies, as states and as nations. As individuals, we make choices based on our knowledge, values and desires. In the democratic societies that we live in, individuals also participate in decision making through contributions of ideas, comments and criticisms. Having the background knowledge to make realistic choices and sensible decisions and/or contribute to discussions about proposed innovations, are good reasons to know some science.

SCIENTIFIC LITERACY

Science encompasses both explanatory beliefs and pragmatic knowledge that are systematised, dynamic and cumulative in a way which can be shared. But, unlike the 'common sense' approach to understanding the world, scientific understanding is not something that most will acquire just through 'growing up' or through the informal processes of learning. Science learning and ultimately a nation's scientific literacy will depend on schools, although not exclusively. Other non-formal science learning outlets such as the Internet, media sources, science discovery centres, museums and extra-curricular programs (for example, science clubs and science competitions) are all valuable components in developing an individual's scientific literacy.

Carl Sagan (1997, p. 26), astronomer, science author and television presenter of the series *Cosmos* wrote:

> We've arranged a global civilization in which the most crucial elements -- transportation, communications, and all other industries; agriculture, medicine, education, entertainment, protecting the environment; and even the key democratic institution of voting -- profoundly depend on science and technology. We have also arranged things so that almost no one understands science and technology. This is a prescription for disaster. We might get away with it for a while, but sooner or later this combustible mixture of ignorance and power is going to blow up in our faces.

The arrangement so that 'almost no one understands' is not done intentionally but by the way the channels of information and learning have evolved in our society. These channels include the various media outlets (TV, newspaper, radio, videos, Internet), private and public discussions among people and/or courses undertaken through the formal channels of education in schools, colleges and universities. It is important to see the formal teaching of science happening within this wider context where media reports and general discussions in the community (for example, blogs and newspaper comments online) are embraced within the teaching. As more scientific and technological issues dominate public debates at national and international levels, teaching students to be scientifically literate would prepare them to be people who are both participatory (in debates and discussions) and informed (or know where to find information) about these issues.

The term scientific literacy is increasingly becoming more widely used to refer to the general public understanding of science. The term has caused some confusion. It does not mean just knowing about the vocabulary, language and writing style used in science – these being the basic functional meaning of 'literacy'. It is used as a term to describe an approach to science education where it assumes a functional level of knowledge of terms and the familiarity with writing style, but puts emphasis on the individual's capacity to be comfortable with thinking about and discussing scientific issues, with functional literacy as one of the elements in the development of this capacity. The notion of scientific literacy and its evolution over the last few decades will be discussed in more detail in Chapter 2.

Issues in Science Education

Research in the past decade has indicated a number of challenges for science education. In the US, a 2007 *Congressional Research Services* report indicated these challenges as "school curriculum and the quality of science instruction, student interest in science, the shortage of qualified teachers, teacher training and retraining, student achievement on science and mathematics measures, and the participation of minorities and women in science" (Matthews, 2007, p. 3). Similar issues such as the decline in uptake of science at sixth form and in undergraduate courses, the quality of science education and shortage of qualified teachers have been reported in the UK (Smith, 2010). In Australia, the research has shown that student disengagement with science begins in the early years of secondary education, resulting in a decline in student participation at upper secondary and post-compulsory science education (Dekkers & De Laeter, 2001; Goodrum, Hackling, & Rennie, 2001; Tytler, 2007). The OECD[1] *Programme for International Student Assessment* (PISA) 2006 showed that many students in the country continued to perform below the OECD baseline in science and mathematics, in particular Indigenous students and students from communities of low socio-economic status (ACER, 2007).

The decline in student interest has been attributed to traditional science curriculum being uninteresting and irrelevant (Fensham, 2004; Goodrum et al., 2001; Lyons, 2006). Many school students find learning science concepts difficult and lack confidence in their own abilities (Goodrum et al., 2001; Parliament of Victoria Education and Training Committee, 2006; Schreiner & Sjøberg, 2004). Contributing to the disengagement are factors such as

[1] Organisation for Economic Cooperation and Development

textbook-based work, repetitive exercises, memorisation of facts, formulas and procedures and a lack of connection between classroom content and the real-world context. These factors leave gaps in the students' knowledge and understanding of science. One of the means of overcoming the gaps is to capitalise on the affordances digital technologies offer in enabling the learning of science (Webb, 2005). The motivating influence of technology on young people's learning has been well documented (for example, Crook, Harrison, Farrington-Flint, Tomás and Underwood, 2010; Higgins, 2003; Keogh, 2011; Ng, 2008; Ng & Gunstone, 2002; Pittard, Bannister & Dunn, 2003; Wallace 2002) and should be capitalized upon to support the development of scientific literacy using digital technologies.

ENGAGING YOUNG PEOPLE TO DEVELOP SCIENTIFIC LITERACY WITH DIGITAL TECHNOLOGIES

The current generation of students and undergraduates in our schools and tertiary institutions has been referred to as 'digital natives' (Prensky 2001) or the 'millennials' (McMahon & Pospisil, 2005; Oblinger, 2003; Shepard, 2004). They are students born in or after 1980. This generation of students has grown up in an immersive computing environment and has been described as young people who are so attached to technology that life without their mobile phones, iPods, computers or being online is unimaginable (Chansanchai, 2006). Similar findings from the *Digital Youth Project* (Ito & team, 2008) indicated that "social network and video-sharing sites, online games, and gadgets such as iPods and mobile phones are now fixtures of youth culture" (p. 1). The conceptual shift over the last two decade from Information Technology (IT) to Information Communication Technology (ICT) reflects a shift from IT as providing, storing and processing information to a focus on ICT's ability to "strengthen and multiply communicative relationships – between ideas, especially as a feature of non-linearity and hypertext, as well as between people" (Nixon, Atkinson and Beavis, 2006, p.133). Consistent with this, the young people of today are comfortable with using digital technologies, embracing them in their everyday lives and are frequently online users, networked socially through technology to stay in contact with friends, family and peers via email, mobile phones, SMS, MSN, video conferencing, discussion boards and chatrooms. They become part of Web 2.0's online communities and collaborate to discuss a topic or an issue (for example through blogs, *Skype* or *VoiceThread)* or create and share information (for example through wikis, *Slideshare* and *YouTube*). This 'multiplicity of communicative relationships' takes place formally and informally wherever the Internet is accessible. In this respect, mobile devices further enhance the communicative capabilities and access to information through its inherent ubiquitous characteristics. In addition to content and pedagogy in curriculum design, the view that learning is a cultural activity needs to be recognized. By linking students' learning experiences outside of educational institutions to those within the institutions' contexts, students' learning could be enhanced and institutions should be supportive of this.

Chapters in This Book

The technologies mentioned above are enabling technologies for developing scientific literacy. These and other digital technologies for science learning will be elaborated in Chapters 4, 5 and 6. The use of digital technologies needs some level of digital literacy. The book attempts to argue, from Chapter 3 onwards, the importance of being digitally literate to use technology effectively, in this case for developing scientific literacy. A conceptual framework for digital literacy is presented in Chapter 3. The chapter discusses the complexity of digital literacy with its cognitive, technical and social-emotional dimensions and the multiple literacies within them. One of the multiple literacies in the digital literacy framework is 'multiliteracies' (New London Group, 1996). The multimodal perspective of multiliteracies for enhancing science learning is discussed in Chapter 4. The chapter explains multimodality, its importance for science learning and provides examples of using different modes of representations for the development of scientific literary. In Chapter 5, the justification for the title of this book is made, explaining why students who are digitally literate and multiliterate are empowered to learn science better to develop their scientific literacy. A conceptual framework demonstrating the relationships between the various literacies (scientific literacy, digital literacy and multiliteracies) is shown in the chapter. Chapter 6 looks at the pedagogy of using multimodal means of teaching and learning science, with the provision of some pedagogical examples. Chapter 7 discusses the implications of using digital technologies in science teaching, focusing on the need for teachers to be digitally literate as well as understanding about multimodality in their teaching of science. Chapter 8 is the concluding chapter for the book and summaries the ideas presented in the book. It discusses some of the educational trends that are developing resulting from the availability and advancement of educational digital technologies such as the one-to-one technology-enabled learning concept, learning spaces, Science 2.0 and impact of technologies on the power-relationship between teachers and students.

Chapter 2

SCIENTIFIC LITERACY

INTRODUCTION

This chapter defines scientific literacy and introduces the concept and its development at two levels, a specialist level and a generalist level. It will discuss the importance of being scientifically literate and the skills and knowledge associated with it. Based on the literature and other reports around the world, the chapter will present the theoretical dimensions and frameworks for scientific literacy proposed by researchers and establish a position for the purpose of this book. It will look beyond the classroom development of scientific literacy and discuss informal learning that takes place outside the classroom, particularly learning that is facilitated by digital technologies.

SCIENTIFIC LITERACY

The term 'scientific literacy' was first used by Paul DeHart Hurd in 1958 in his article entitled: *Science Literacy: Its Meaning for American Schools*. The term has become a key concept in school science curricula and standards in many countries around the world. Its definition and meaning has however caused some confusion. Its meaning goes beyond the basic functional meaning of 'literacy' that is the knowing about science terminologies and the writing style for scientific reports. Widely used as a term to refer to concerns about the general public's understanding of science, the term is used to describe an approach to science education that assumes a functional level of knowledge of scientific terminology and a familiarity with its writing style (Hodson, 2008).

Hodson (2008) further added that scientific literacy also presupposes some basic knowledge of mathematics and the understanding of the dependence of science on mathematics.

The approach mentioned above refers to an individual's capacity to think critically and discuss science-related societal issues that arise out of science and technology development.

Millar (1996) has looked at the range of views given for "why should we promote the public understanding of science?" and found rationales grouped into five categories:

1. the *economic* argument: that there is a connection between a nation's wealth and the level of its public understanding of science.
2. the *utility* argument: that understanding of science is helpful in everyday life
3. the *democratic* argument: that understanding of science assists citizens to take part in societal decision making
4. the *social* argument: that public understanding of science maintains links between science and the wider culture and
5. the *cultural* argument: that understanding of science should be encouraged as science is the major achievement of our culture.

Similar rationales were echoed by The Royal Society (1985, p. 9, cited in Hodson, 2008) whose assertions indicated that:

> Improving the public understanding of science is an investment in the future; not a luxury to be indulged in if and when resources allow...(it) can be a major element in promoting national prosperity, in raising the quality of public and private decision making and in enriching the life of the individual.

Both Millar and The Royal Society's arguments serve to distinguish between those who see scientific literacy as the possession of knowledge and skills required to embark on a professional career as a scientist or engineer and those who see it as the capacity to understand and participate intelligently in discussions about the impact of science and technological advancements on one's own life and on society in general. This distinction is made more explicitly by the European White Paper on Education and Training (1995, p.11) which, in making a case for scientific literacy asserted that:

> '..this does not mean turning everyone into a scientific expert, but enabling them to fulfill an enlightened role in making choices which affect their environment and to understand in broad terms the social implications of debates between experts. There is similarly a need to make everyone capable of making considered decisions as consumers.'

This distinction is important as scientific literacy is widely advocated not just as a component of science curricula but often as the fundamental purpose of science education during the compulsory years of schooling. Bybee and DeBoer (1994) who conducted a historical review of the goals for science curriculum have summarised them as being to assist students to acquire scientific knowledge, to learn the processes or methodologies of the sciences and to understand the applications of science, especially the relationships between science and society and science-technology-society.

At the simpler level of argument, these goals could be seen as (i) focusing on producing skilled science and engineering students to meet the scientific/technological needs of society and (ii) giving all students some knowledge of science as part of their general education (Fensham, 1985). The latter provides the basis for developing scientific literacy that is akin to the public understanding of science and a view of science education as being broader than learning science facts and concepts in preparation for employment or further study. In Australia, the 2000 national report *The Status and Quality of Teaching and Learning of Science in Australian Schools* (Goodrum, Hackling & Rennie, 2001, p. ix) stated that:

> The purpose of science education is to develop scientific literacy which is a high priority for all citizens, helping them to be interested in, and understand the world around them, to engage in the discourses of and about science, to be skeptical and questioning of claims made by others about scientific matters, to be able to identify questions and draw evidence-based conclusions, and to make informed decisions about the environment and their own health and well-being.

In the US, the *Science for All Americans*[1] document, in setting out the rationale for the major long-term science reform initiative *Project 2061* defined the scientifically literate person as:

---

[1] Science for All Americans is a result of Project 2061, a long-term AAAS initiative for the advancement of literacy in science, mathematics and technology. It consists of a set of recommendations on the essential understandings and ways of thinking for all citizens in a

> one who is aware that science, mathematics, and technology are interdependent human enterprises with strengths and limitations; understands key concepts and principles of science; is familiar with the natural world and recognizes both its diversity and unity; and uses scientific knowledge and scientific ways of thinking for individual and social purposes
> (AAAS[2], 1989; p.ix-x).

At an international level, the OECD[3]'s *Programme for the International Student Assessment* (PISA), which is an internationally set standardised assessment of literacy, mathematics and science for 15-year-old school students, has an operational definition for the term scientific literacy as:

> The capacity to use scientific knowledge, to identify questions and to draw evidence-based conclusions in order to understand and help make decisions about the natural world and the changes made to it through human activity.
> (OEDC/PISA, 2003; p.133)

While the wordings for the various definitions stated above vary somewhat, they all emphasise that scientifically literate individuals should show the ability to apply scientific, mathematical and technological knowledge and skills to aspects of their own lives and to be active participants of a society where science-related issues and discourses prevail. While not stressing that scientific literacy is for the training of future scientists and engineers, the definitions imply the need for students to develop an understanding of the nature of science. This includes the learning of science concepts and those related to the scientific method in school-based curriculum.

The literature also contains other expanded views that are proposed by other organisations and researchers of what a scientifically literate person is. Norris and Phillips (2003, p. 225) have summarised these from various research publications, stating that scientific literacy is used variously in one or more of the ways stated in the left hand column of Table 1. On the right hand column of Table 1, I have grouped Norris and Phillips' list into 4 categories. The first two categories focus on science content knowledge/understanding

world shaped by science and technology. Available online at http://www.project2061.org/publications/sfaa/default.htm

[2] American Association for the Advancement of Science

[3] Organisation for Economic Co-operation and Development

and the ability to solve problems with it. The third category relates to public understanding of science - the ability to read science-related media articles and participate in intelligent and informed discourses of science-based issues. In order to do so, the individual needs to exercise critical thinking skills to have a balanced view of the issues at hand. The fourth category is a more personal but important aspect of scientific literacy where the individual's affective and aesthetic development would ensure the development of a liking towards science and the lifelong pursuit of scientific literacy. The aesthetic dimension of being scientifically literate has also been proposed by Hazen (2002), as involving a deeper appreciation of everyday activities and living through the understanding of science.

Table 1. Categorisation of the characteristics of Norris & Phillips' (2003) scientific literacy

Scientific literacy (Norris & Phillips, 2003, p. 225)	**Categorisation of Norris & Phillips' list**
Understanding of basic scientific ideas	Content knowledge acquisition: Developing science knowledge and understanding
Understanding of the nature of science	
Knowledge of what counts as science; the ability to distinguish science from non-science	
Understanding science and its applications	Problem solving: Applications of science to solve problems in society
Ability to use scientific knowledge in problem-solving	
Knowledge needed for intelligent participation in science-based social issues	Public understanding of science: Active participation in discourses relating to science-based societal issues that affect the individual's personal life and the environment
Knowledge of the risks and benefits of science	
Ability to think critically about science and to deal with scientific expertise	
Ability to read with understanding articles about science in popular press	
Appreciation of and comfort with science, including a sense of wonder and curiosity	Affective and aesthetic appreciation : Personal development for lifelong learning of science
Ability and wish to be an independent, lifelong science learner	

It should be noted that the list of Norris and Phillips' (2003) description of a scientifically literate person captures the descriptions of a scientifically literate person proposed by Goodrum et al. (2001), AAAS and OECD presented above. There are however, still other dimensions of scientific literacy proposed by researchers in the literature such as the association with technological literacy and the historical and philosophical perspectives of scientific literacy.

OTHER DIMENSIONS OF SCIENTIFIC LITERACY

In his chapter *STS Education: A Rose by Any Other Name,* Glen Aikenhead (2003) wrote a comprehensive account of the emergence and evolution of the Science, Technology and Society (STS) reform effort that began in the 1970s. The STS framework sought to relate science to other discipline areas of the curriculum, that is, relating science to technology and science to society. STS is defined by the National Science Teachers Association (NSTA, 1990-91) as science and technology education in the context of human experiences and that its approach promotes interdisciplinary and cross-disciplinary learning in the school curriculum. In summarising the publications on the position statements of the National Science Board's (NSB, 1983) *Educating Americans for the 21st Century* and NSTA's *Science-Technology-Society: Education for the 1980s*, Bybee, Powell, Ellis, Giese, Parisi and Singleton (1991) asserted that students need to understand both the history and nature of science and technology as part of the STS curriculum. The NSB's (1983) emphasis on science and technology in a social context were provided with reasons that:

- science and technology are integral parts of today's world
- technology will continue to change our society
- complex social issues have emerged as a result of using science to solve problems
- students, as future citizens, need to be prepared to critically address these issues through an understanding of the impact of how science and technology impact on the quality of life

while NSTA's (1982) position statement in *Science-Technology-Society: Education for the 1980s* listed the characteristics of a scientifically literate person as being able to

- understand the interrelationships between science, technology and society (how society influences science and technology, and how science and technology influence society)

- understand that scientific knowledge is generated through the inquiry process and in formulating theories
- recognise the history of science and its dynamic nature and that its theories are subject to change as evidence accumulates, due in part to technological advancement enabling better and more powerful equipment to be constructed for science pursuit
- understand the benefits and limitations of science and technology and
- use scientific knowledge in decision making.

Hence, in addition to the dimensions that are described in the previous sections of this chapter, the STS model for scientific literacy entails an understanding of science and technology and their relationships to society, the history of science and the benefits and limitations of science and technology. Implications of NSTA's STS approach for science teachers' professional development included teachers being introduced to constructivist instruction in a STS context, the integration of concepts from other science disciplines and the use of real-life contexts in their instructions.

On the other side of the argument, there were critics of the science-technology-society movement who feared that science would lose out to technological issues and social analysis, leading to a lack of attention to the science basics (DeBoer, 2000). Bybee (1997) was also critical of the changes made by science educators to science programs and practices without a shared vision of what they were trying to accomplish. International reform documents produced around 1989 and beyond had therefore attempted to describe scientific literacy in terms of science content, science processes and social contexts (AAAS, 1989; DfES, 1995; AAAS, 1993; Curriculum Corporation, 1994; National Research Council, 1996).

FRAMEWORKS OF SCIENTIFIC LITERACY

Researchers have attempted to break the complexity of defining scientific literacy by proposing different frameworks for the concept. Shen (1975) suggested that there are three components for scientific literacy, these being practical, cultural and civic. The practical aspect of Shen's scientific literacy is based on science being an integral part of everyone's life, and that science knowledge and skills provide practical assistance in helping people make informed decisions and choices of the way of life that are best suited for them.

The civic aspect enables citizens to be able to read and understand reports and articles in the media, to engage in public conversations about the ethical, moral and social issues of scientific discoveries and to have an influence over governments' decision making. The cultural aspect of Shen's scientific literacy is about the importance of learning science in its social and human context and knowing about the science ideas that brings about major cultural achievements.

In *The Myth of Scientific Literacy,* Shamos (1995) was critical of the broad definitions of scientific literacy indicating that they were written by so many people that it contained 'virtually all of the objectives of science education that have been identified over the years' (p. 590, cited in Laugksch, 2000) that were open to interpretation, assumption and how to achieve them. Shamos proposed a framework of scientific literacy that is hierarchical comprising of three forms: cultural, functional and 'true'. Cultural scientific literacy is the lowest form of scientific literacy with the individual having a general level of science knowledge that enables him/her to read newspaper articles and follow science-related debate in the media. Functional scientific literacy requires the individual to have a command of science vocabulary to "converse, read, and write coherently, using such science terms in a perhaps non-technical but nevertheless meaningful context" (p. 8). 'True' scientific literacy is the most difficult level to attain and is characterized by scientific habits of mind such as logical reasoning, the ability to think critically, the role of experiments and reliance on evidence and other elements of scientific investigations. Shamos argued that it is unlikely that students will be able to achieve the 'true' form of scientific literacy and that they can only achieve functional scientific literacy at best. Hence, he proposed that school educators should teach for scientific awareness rather than scientific literacy where the latter requires the grasping of the abstractions of science (DeBoer, 2000).

Scientific literacy frameworks similar to Shen (1975) and Shamos' (1995) have also been proposed by other researchers, for example Bybee (1997), Pella, O'Hearn and Gale (1966), Branscomb (1981) and Miller (1983). Bybee (1997), drawing on the US *Science Education Standards* (National Research Council, 1996) and *Benchmarks* (AAAS, 1993), constructed a framework containing several dimensions of scientific literacy, these being nominal, functional, conceptual and procedural, and multidimensional scientific literacy. An individual possessing

- nominal scientific literacy, the lowest form of scientific literacy, is able to recognize a science-related concept but the level of understanding is low and often misconceived
- functional scientific literacy is able to describe a science concept using scientific vocabulary but with limited understanding of it
- conceptual and procedural scientific literacy is able to develop some understanding of the conceptual schemes of science, procedural knowledge and skills, relationships between the parts of a science discipline and the conceptual structure of the discipline
- multidimensional scientific literacy understands the distinctions between science and other discipline areas, knows about the history and nature of science disciplines and understands science in a social context.

Koballa, Kemp and Evans (1997) in comparing Bybee (1997) and Shamos' (1995) forms of scientific literacy have aligned them as follows:

- Bybee's nominal and functional dimensions with Shamos' cultural scientific literacy
- Bybee's conceptual and procedural dimensions with some aspects of Shamos' 'true' scientific literacy and
- Bybee's multidimensional dimension with Shamos' 'true' scientific literacy. Incorporated into the multidimensional and 'true' scientific literacies is the understanding that apart from knowing the content (science concepts and procedures), knowledge of the history and nature of science and an understanding of the influence of human culture on science and science's influence on human culture and the scientific enterprise need to be developed.

Like Shamos (1995) who do not think that many individuals will be able to achieve 'true' scientific literacy, Bybee (1997) asserted that it is impossible to achieve multidimensional scientific literacy in all scientific domains. He argued that it is possible to be highly literate and develop expertise in one area, without being career-oriented. For example, a person who is passionate about building electronic gadgets as a hobby could attain high literacy in physics but low literacy in other scientific domains. Hazen (2002) also made similar assertions in that scientists with expertise in a particular field often lack scientific literacy characteristics. He asserted that it is not uncommon to find physicists being uninformed in areas outside their field of expertise, such as lacking in the broad knowledge of biological sciences. The converse of biologists lacking in the broad knowledge of the principles of the physical sciences is also true. Hazen summarised scientific literacy as simply a mix of concepts, history, and philosophy that help an individual understand the scientific issues of his/her time. He stated that a scientifically literate person has a broad understanding of the most general principles of science and possesses within his/her knowledge sufficient facts and vocabulary to understand scientific issues in print and electronic media with similar ease as understanding news in politics, sports or the arts. He argued for the importance of scientific literacy from the aesthetic, civics and intellectual coherence perspectives. The aesthetic dimension of being scientifically literate involves individuals appreciating science through their everyday activities. From the civics perspective, he argued that the general welfare of a nation is stronger if its citizens are scientifically informed. From the intellectual coherence perspective, he argued that because our society is so inextricably linked to science discoveries that they often play an important role in setting the intellectual climate of periods in human history. For example, nano-science and technology, hailed by some as the fifth industrial revolution (Treder, 2004) is currently one of the dominating topics of research and discussions of science.

THE COMPLEXITY OF DEFINING SCIENTIFIC LITERACY

Up to this point of this chapter on scientific literacy, it is evident that the concept is conceived broadly by researchers, sometimes with the use of similar terminologies to mean different things. For example, Shamos' (1995) cultural scientific literacy is the lowest in the hierarchy of his framework but appears to be the highest in Shen's (1975) framework. Shamos' functional scientific literacy would reflect Shen's practical scientific literacy. My review of the literature is far from exhaustive but demonstrates the complexity in defining the concept of scientific literacy. Since this book is aimed at scientific literacy largely at the school level, it is appropriate to look at the positions of scientific literacy in school curricula. Connor (2007) has analysed the science curriculum document of several countries including Australia, Korea, Norway and New Zealand, for their explicit attention to scientific literacy in the rationale/goals and/or the content sections of the respective country's science curriculum documents. A meta-analysis of the paper with additional inclusion of the national science curricula of Australia and UK is shown in Table 2. In the table, the ticks represent explicitly stated and/or implied dimensions of scientific literacy from the nation's science curriculum document. The dimensions were identified with related words, for example,

- Nature of science – scientific method and thinking; evidence-based
- Societal Issues – ethics, citizenship
- Personal development for lifelong science learning – positive attitudes, curiosity, persistence
- Career – science-related career or as a profession

The table shows that 'nature of science' and 'applications of science' are scientific literacy dimensions in all the school science curriculum of these nations. The explicit relating of science to technology or to the environment is not evident in some curricula (although it does not mean that they are not taught) while history of science is explicitly evident only in the Australian curriculum.

Table 2. Indicated dimensions of scientific literacy in schools' science curricula

Country	Nature of Science	History of Science	Applic'ns of Science	Societal Issues	Science and Technology	Science and Env'ment	Personal dev'pment for lifelong learning	Career
Australia	✓	✓	✓	✓			✓	✓
UK	✓		✓	✓	✓		✓	
*Korea	✓		✓		✓		✓	
*Norway	✓		✓	✓	✓	✓	✓	
*NZ	✓		✓	✓	✓	✓		

*Data from Connor (2007)

For the purpose of this book, I will attempt to consolidate the views on scientific literacy discussed thus far in order to identify those dimensions that could be developed with the support of information communication technology (ICT). The key elements summarized below indicate that a scientifically literate individual should:

i. develop a knowledge of the general principles of science and the processes of how scientific theories are generated and their applications for solving problems in society

ii. understand the applications of scientific theories, principles and methods to solve problems in society, including understanding of the strengths and limitations of scientific theories, questioning the integrity of evidence shown and any ethical or social concerns that may be associated with it
iii. be able to read and critique media reports, including online reports, and if necessary to undertake further research and analysis of the topic
iv. be able to collaborate and communicate with peers and experts (for example via emails, blogs or video-conferencing) and the general public (for example through blogs or making comment in online newspaper, journals or magazines) about science-related issues
v. develop an awareness that technology plays a major role in the advancement of science either by supporting direct discovery through instrumentations and for communication and collaborative purposes and
vi. develop curiosity, persistence and positive attitudes towards lifelong learning of science.

Nbina and Obomanu (2010), in their review of the literature on scientific literacy, have suggested that the trend in defining the concept seems to be moving away from the short term product approach, that is, a focus on knowing about pure science skills and abstract ideas, towards what Shamos (1995) would classify as 'science awareness' where the emphasis is on the ability to make decisions related to science-based issues in everyday living and to participate as effective citizens in a science and technologically oriented society. However, within the context of this book where the use of ICT is the underlying theme for assisting students in their development of scientific literacy, there will be emphasis on the study of science content and concepts as this is a necessary and crucial component in the development of scientific literacy. Developing scientific literacy could be seen as a continuum of learning principles and concepts of science and their applications in real-life situations. Otherwise, the approach will be taking a "back-to-front" approach as argued by Sir Richard Sykes, Rector of the Imperial College of London and the former chairman of the pharmaceutical giant GlaxoSmithKline, whose comments on the new science curriculum were:

> a science curriculum based on encouraging students to debate science in the news was taking a "back-to-front approach and that before the future citizen can contribute to the decision-making process, they need to have a

> good grounding in the fundamentals of science and technology, rather than the soundbite science that state school curriculums are increasingly moving towards. (cited in Braithwaite, 2006, online)

In adopting ICT for learning and arguing for how it can empower scientific literacy, the six key elements of a scientifically literate person that I have listed above could be further grouped into 3 categories:

1. *Developing science knowledge and skills and their applications to solve problems in society.* This category embraces element (i) which gives attention to learning science content, concepts and processes to prepare students to develop scientific literacy at two levels – generalist and specialist levels. At the generalist level, this category equips students with the understanding of basic science principles to prepare them for functional and participatory citizenships. At the specialist level, learning science content at a higher level such as the undertaking of an undergraduate science degree will prepare students for the pursuit of science and engineering careers in the future. ICT has the potential to cater for the needs of students with different interests and abilities.
2. *Public understanding of science: application of science knowledge in issues associated with science/technology in society.* Elements (ii) to (v) forms this category that shapes the widely acknowledged notion of scientific literacy, that is, the preparation of students to understand the impact of science and technology on their own lives and on society, and to be active citizens of society by contributing voices/opinions to government undertakings and policies. ICT provides opportunities to learn in context through dealing with real time data and information or simulated scenarios that fosters development of scientific literacy through inquiry-based learning.
3. *Affective and aesthetic dimensions of scientific literacy: personal development for lifelong learning of science.* Element (vi) is a category that relates to the individual developing a personal interest in societal issues and understanding of how the development of science and technology will continue to impact on his/her own life throughout his/her lifetime. In this respect, it is desirable to culture the attribute of curiosity that would include being skeptical, asking questions and seeking for answers, as part of the development of scientific literacy. The vast amount of information and resources on the Internet provides

opportunities to develop this aspect of an individual's scientific literacy, as well as enabling its development through self-directed and lifelong learning.

Informal Scientific Literacy Development

Most of the literature on scientific literacy has focused on formal education, that is, what the curriculum should look like to teach scientific literacy and how teachers should be teaching the curriculum. Dierking, Falk, Rennie, Anderson and Ellenbogen (2003), however asserted that learning science is dynamic and cumulative, emerging over time through a diverse range of experiences that are motivated by an intrinsic desire to learn. The range of experiences includes developing scientific literacy in non-formal and informal learning environments. These environments have been categorised by Honeyman (1998) and Stocklmayer, Rennie and Gilbert (2010) as (i) museums, science centres such as Scienceworks, Exploratorium, Discovery Centres and similar institutions (ii) community programs such as industry, university and after-school science-based programs and (iii) the media reporting science-related news.

Informal and Non-Formal Education of Science

The terms 'informal' and 'non-formal' are often used interchangeably in the literature and in everyday conversations. Coombs and Ahmed (1974) were one of the first authors to make the distinction between formal, informal and non-formal education in the literature. Little variations to this distinction appear to have been made over the years (for example see Eshach, 2007; Rogers, 2005). The general consensus is that formal education is highly institutionalized, chronologically and hierarchically structured and motivation is typically more extrinsic. Informal learning is everywhere, spontaneous, highly contextualised and motivation is more typically intrinsic. Non-formal learning falls in between these two extremes in the continuum of education. It is learning that occurs in an organised but highly adaptable manner in institutions outside the formal system of education. Motivation could be extrinsic but more typically intrinsic. As non-formal and informal learning is highly participatory learning, Rogers (2005) included non-formal learning in his continuum of learning as shown in Figure 1. According to this description, science centres, museums and community-based science programs would fall into the non-formal learning environment. Informal science learning would include casual conversations with people outside of school and obtaining information from the media – newspapers, the Internet, magazines, journals and television.

formal education **non-formal education** **informal education**

Figure 1. Continuum of learning.

In many non-formal educational environments such as museums, scientific literacy goals are being considered as they work to support reform in schools. For example, Boston's Museum of Science used *Project 2061*'s principles for effective science learning and teaching to design a series of interactive exhibits. The Cranbrook Institute of Science in Bloomfield Hills, Michigan, redesigned its exhibits based on *Science for All Americans'* description of systems and exhibits at *Science Alive! (*an interactive science centre in Grand Rapids, Michigan) focused on four chapters from *Science for All Americans*. Hence non-formal science centres such as these, have the potential to fulfil a major role for science education and in the promotion of scientific literacy (Rennie & Williams, 2002). These centres are intrinsically motivating and engaging through building curiosity in students (and the general public), allowing for the making of observations and conducting of science-related activities (Ramey-Gassert, Walberg & Walberg, 1994). They present science content that students can relate to personally, making it more appealing to them over the content studied in classrooms which is often divorced from their real world. Consequently science centres have the potential to provide students with opportunities to be engaged in immersive learning. Immersion in turn increases students' motivation and engagement with the learning, leading to higher levels of satisfaction and achievement in the understanding of science concepts (Dede, Salzman, Bowen-Loftin & Sprague, 1999; Raja, Bowman, Lucas & North, 2004). Furthermore, the psychological impact of 'presence' - the sense of being there (Freeman, Lessiter & IJsselsteijn, 2001; Lombard & Ditton, 1997; Yu, 2005) is known to correlate with increased attention and increased effectiveness of the content being learned (Darken, Bernatovich, Lawson & Peterson, 1999).

As many of the display, both static and interactive, in science centres are in digital form, being digitally literate would help the students to interact effectively with the digital resources for their learning. Other ad hoc, contextual informal learning of science would be from information obtained from the Internet's interactive sites, or through conversations via Web 2.0's social networked communities. The advantages of being digitally literate in helping students learn science in digitally enriched environments are discussed in the next chapter.

Self-Directed and Lifelong Learning

John Dewey (1938) said that "The thirst for learning is the most important attitude that can be formed" (cited in Karsenti, Brodeur, Deaudelin, Larose, & Tardif, 2002, p. 1). Hence, inculcating the desire to keep learning should be a key aim of education. Beyond the formal educational years, scientific literacy will continue to develop through informal, self-directed learning at home or at work (Drucker, 1994; Sargant, 1991). In this respect, Lave & Wenger (1991) contended that lifelong learning becomes intertwined activities where work, leisure, learning and collaboration within communities are situated within contexts. Educational institutions should prepare their students for lifelong learning by providing opportunities to develop their critical thinking skills and other lifelong learning skills such as independence and responsibility that will enable them to learn in different settings throughout their private and work life after formal education. Being able to self-direct their own learning is one of the major steps in achieving this goal, as Fisher & Scharff (1998, online) stated:

> A lifelong perspective implies that schools and universities need to prepare learners to engage in self-directed learning processes because this is what they will have to do in their professional and private lives outside of the classroom.

A definition of self-directed learning is the development of autonomous learners who are able to take control of and responsibility for their own learning, including knowing where to seek for help when difficulties are encountered (Garrison, 1992; Tough, 1967). Smith (1996) proposed two characteristics of self-directed learning as (i) a continuous activity where the individual has 'authentic control' and decides on all aspects of the learning such as planning, doing and evaluating the learning and (ii) the ability to access and select from a range of appropriate resources. Hence in educating students to be self-directed learners, and to cater for differences in background and expertise, there need to be flexibility in both the diversity of resources offered and freedom for the learners to select learning materials (Sinitsa, 2000). ICT offers affordances that enable students to develop as self-directed learners, for example, interactive resources such as drill and practice that provide instant feedback. Other skills that need to be promoted for self-directed learning are critical thinking skills (for example seeking, selecting and evaluating appropriate resources from the Web), social and teamwork skills. The latter two are to assist with learning through collaborative and collective efforts that are facilitated by Web 2.0 technologies.

Useful devices for promoting self-directed, lifelong learning are mobile technologies. These technologies offer greater freedom to learn ubiquitously (Waycott, Jones, & Scanlon, 2005). They allow the learner to be mobile and to interact with a variety of contexts so that learning can take place both formally and informally. They offer the potential for ubiquitous learning in ways that are independent and self-paced. Learning with mobile devices is personalised so that it fits an individual's learning path and their social networks. A personal learning environment that uses technology, such as a personal digital assistance (PDA) or smartphone, as a ubiquitous learning tool, recognizes that learning is continuous and not only made available by a single learning provider such as a school. Teaching students to be digitally literate to learn independently with ICT is a step towards preparing them to be self-directed, life-long learners. Digitally literate individuals should be able to make use of ICT to assist in curiosity-building and solution seeking to further develop their scientific literacy. In order to prepare students to make use of ICT for self-directed and lifelong learning, schools should make explicit the teaching of digital literacy and how to maximize the potential of various digital devices and the resources available to these devices for learning. In addition, teachers should also develop meaningful tasks that bridge school-home (that is, formal-informal) learning.

CONCLUSION

The development of scientific literacy in students is an essential responsibility of educational institutions and their teachers. Students who are scientifically literate would have sufficient science knowledge, or know how to obtain science information from available resources to assist them to understand the science-related events that happen around them. For example, the introduction of a carbon tax in Australia has generated a lot of debate between people from all walks of life. A scientifically literate individual should be able to join in the discussions and critically assess both sides of the debate to come to his/her own conclusion of the value of the tax. The individual should be able to listen to or read up on the evidence about global warming, including the limitations of these data, to decide on the degree of impact that carbon emissions are having on the environment.

Developing scientifically literate students is also about preparing the next generation of scientists and engineers. Hence teaching science to both the to-be scientists/ engineers and those not pursuing such careers would mean being able to differentiate the science curriculum and apply pedagogy that engages both groups of students to learn. ICT has the potential to cater for students needing to learn more content. It also has the potential to assist weaker students to learn science content through the different formats of presentations enabled by ICT. Learning with ICT would require students to have some level of digital literacy, the levels being dependent on the complexity of the technology used and the task being undertaken. The concept of digital literacy and its importance for developing scientific literacy are discussed in the next chapter.

Chapter 3

DIGITAL LITERACY: A GENERAL PERSPECTIVE

> I have noticed that the students are spending a lot of time on the technical and presentation part of the assignment, are they really learning? (*pre-service teacher, UNSW, Australia, on integrating ICT into science teaching*)

This pre-service teacher's voice is not alone. The issue is one of the concerns that teachers have to deal with when they integrate ICT into their teaching and their students' learning. It is a legitimate concern because when students pay more attention to the technical aspects of the software, the focus on the discipline's content and its learning could diminish. This chapter will discuss a general perspective of digital literacy and propose that it should be explicitly taught and developed in students so that the focus shifts back to the learning of the content. The chapter will put forward a framework for digital literacy and discuss the complexity of the skills and cognitive abilities that need to be developed to be digitally literate.

INTRODUCTION

Similar to the argument for being scientifically literate is the argument that students and the general public should be digitally literate. This is because digital technology tools are advancing and proliferating the marketplace at an increasing pace. Their presence is everywhere in our daily lives and they are becoming more central to the individual's learning and social wellbeing. They are also central to the economic development and advancement of businesses, corporations and nations. E-Skills UK (2009) reported on the increasing number of graduates directly employed in 'digital industries', with an estimated 77% of UK jobs involving some level of competency in ICT. The requirement of workers to update skills as technologies get upgraded is a common feature of workplaces in the 21st century. The importance of digital literacy in terms of the economy and employment rates has been stressed by Wynne and Cooper (2007, p. 4):

> At a national level, a growing number of experts predict that a lack of digital literacy will have a dampening impact on economic prospects. Consider that in the next eight years, according to Monthly Labor Review Online (November 2005 p.6) six out of every 10 new jobs will be in professional and service-related occupations requiring, at a minimum, a basic level of proficiency in computers…….. Economic advantage and competitiveness will rest heavily on our ability to equip the 21st century workforce with competitive digital literacy skills.

At the level of the individual, Berson and Berson (2003) noted that the youths of today are accessing a vast amount of information through the various media outlets and are simultaneously creating and disseminating their own messages and creative products through digital technologies. Due to the difficulties in parental and institutional control of young people's access to information from these outlets (for example, television and the Internet), educating them to be digitally literate is the most effective way of safeguarding them from being exposed to harm when being in digitally-enhanced environments. Berson and Berson (2003) also asserted that effective citizenship is derived from a digitally literate population, a similar argument to a scientifically literate citizenship discussed in Chapter 2. Equipping students with scientific and digital literacies in order to prepare them for the 21st century workforce and citizenship are not separate, but intertwined, educational processes. As science education is largely targeted at school aged students and young adults at tertiary institutions, a brief description of the young people of today and how they interact with technology will be discussed next. An understanding of young people's digitally-related activities is essential for teachers if they are to design ICT integrated instructions to engage their students in their learning.

Young People of Today

In 2001, Prensky published a paper where he referred to a generation of people born in and after 1980 as 'digital natives'. He described digital natives as living lives immersed in technology (such as computers, video games, mobile phones and digital music players) and that they learn differently from previous generations of students. According to Prensky, digital natives are active experiential learners, they like receiving information quickly (twitch-speed), are multi-taskers and parallel processers and prefer graphics first over texts. These digital natives have e-lives that revolve around the Internet, for example *eBay*, *Facebook*, wiki, blogs, online games, downloading music etc., where they use technologies for accessing information from the Internet and for interacting with others.

Prensky's concept of digital natives has sparked a few papers arguing against it in recent years (for example Brown & Czerniewicz, 2010; Bennett, Maton, & Kervin, 2008; Helsper & Eynon, 2010; Kennedy, Judd, Churchward, Gray, & Krause, 2008). Some of the digital natives' characteristics that the researchers argued against are:

- *the generation factor where those born after 1980 are digital natives.* The researchers argued that it is not the age that should be considered in describing the youths of today but other more important factors such as the availability of technology and breadth of use, prior experience, self-efficacy and education.
- *the availability of technology to digital natives and their ubiquitous usage.* The researchers argued that the use of technology by young people is different in education in that most lack the skills and strategies to use them for learning.
- *because young people have grown up in a world surrounded by technology, their brains develop differently to the adults of previous generations.* The researchers argued that there is no empirical evidence to suggest that the brain structure is different between adults and those who use the Internet and other technologies frequently.

I think Prensky's generalization of a generation of young people as digital native is no different from society referring to the 'baby boomers' (Jones, 1980) as people born post-world war II between 1946 and 1964. Baby boomers possess certain types of stereotypical characteristics that are often referred to by the media, for example being work-centric, goal oriented and competitive (Kane, nd). There is legitimacy in Prensky's description of the current generation of young people. This generation is born in a digital age, and the digital environment was where they were brought up in and where immersion in digitally-related activities is a common feature. This by definition[1], according to dictionary.com, qualifies them to be called 'natives'. The argument that many digital natives do not use the technology they possess for learning school-based curriculum does not disqualify them from being called digital natives. Young people do not think or know about educational technologies unless they are exposed to them or there is a need to do so. It is the task of educators and teachers to raise awareness of the range of educational technologies that they could use. They need to be taught about these technologies, just like a child born into a community needs to be exposed to or be taught how to use the tools and equipment that are available to the community and other social-cultural dimensions related to the uses. Unlike the learning about social networking and other entertainment tools where it is largely peer-driven and learned through 'tinkering' (Ito et al., 2008), educational technologies would most likely be the last thing in digital natives' minds to seek out, explore and use unless they are introduced to them and only if there is a need to use them. This is the finding of a recent small-scale study that I conducted with 53 undergraduate students studying an *Introduction to eLearning* course. Sixty-four percent of the students (34 students) participated in the research of which 88% of them were in the age group of 18-22 and the other 12% were in the age group of 23-26. The students were introduced to a range of educational technologies such as *Inspiration* for concept mapping, *Prezi* presentation software, *Hot Potatoes* and *SurveyMonkey* for building quizzes, and *Voicethread*, *GoogleSites* and *Wikispaces* for collaborative work. For each piece of software introduced, the students had to construct a meaningful artefact to demonstrate its use, for example they created WebQuests in *Wikispaces* and digital stories using *MovieMaker* or *iMovie* and uploaded them onto *VoiceThread*.

As shown in Table 1, the majority of the 34 students who participated in the research indicated that they knew little or nothing at all about these software (except for *MovieMaker/iMovie*) or concepts such as digital story or WebQuests. The research found that the lecturer or tutor only needed to introduce the unfamiliar technologies briefly by showing some of the features

---

[1] Native = the place or environment in which a person was born (dictionary.com)

of the programs for the vast majority of the students to pick up and use them comfortably and easily. The students were able to learn the newly introduced technologies quite quickly and were successful in constructing meaningful products from the software. The findings could be attributed to the ease with which these students handled technology and the digital literacy that they already possess, having grown up in an environment that is digitally and technologically enhanced.

Table 1. Percentage of students' responses to 'How much do you know about these technologies?" at the start of the course

Technology/Concept	Percentage of students' responses		
	A lot	A little	Nothing at all
Inspiration (for concept maps)	18	27	56
Prezi	12	15	74
Hot Potatoes	0	18	82
Wiki	20	65	15
WebQuest	6	29	65
ePortfolio	12	38	50
Voicethread	6	21	74
Digital Stories	15	27	59
Movie maker; iMovie; Photoshop	50	47	3

Table 2a. Average score of self-rating on a scale of 1-10, for the question 'On a scale of 1-10, I think my level of digital literacy is...'

	Average (N=34)
Start of semester	6.2
End of semester	8.0

In the study, the students were asked to rate their digital literacy on a scale of 1-10 at the start of the semester (before the course) and at the end of the semester (after the course has been completed). The results (see Table 2a) show that the young people generally have quite high self-ratings of themselves as being digitally literate. The average self-rating of the 34 students was 6.2 (out of a maximum of 10) at the start of the semester and this went up to 8.0 at the end of the semester.The result suggests that teaching the students about programs that they were not aware of and providing some initial support on how to use them further improved their self-perceptions of their own digital literacy capacity. Table 2b shows that 85% of the students improved their self-ratings by 1 or more points, while 15% (five students) indicated that the exposure to more technology did not improve their digital literacy. These five students rated themselves very highly on the scale for both the start and end of the semester. There were nine students (27%) who indicated an improvement of 3 points or more. But the majority of the students (20 students or 58%) indicated an improvement of one or two points after being taught a range of educational technologies in the semester.

Table 2b. Frequency and percentage of change in students' self-ratings of digital literacy over the semester

Change in self-rating (*difference between start and end*)	**Change in self-rating of being digitally literate from start of semester to end of semester**							**Total**
	No change	**1**	**2**	**3**	**4**	**5**	**6**	
No. students	5	11	9	5	2	1	1	34
% of students	15%	32%	26%	15%	6%	3%	3%	100%

While the data sample is small and cannot be taken as representative of this age group of students, the results do indicate that the vast majority of students in this study were able to adopt new technologies comfortably and quickly. Observation data for the study also support this conclusion. While the teaching aspect could not be discounted, the finding is likely to be attributed to the ease with which these young students engaged with the technologies. The data also indicate that digital literacy could be taught. Informal interviews with seven students about their definitions of 'digital literacy' indicated that being technically skilled to use the technologies was the most popular response. For example, one of the students proclaimed herself as being "digitally illiterate" as she spent more than two hours figuring out *Wikispaces* and inserting an image into the wiki that she and her partner were constructing for an assignment.

Digital literacy is more than being just technically skilled. For the rest of this chapter, I will be discussing the general meaning of digital literacy, using science-based learning examples where appropriate. In order to establish consistency in meanings with the terms used, I will discuss first what 'digital technologies' mean in this book and its relationship with ICT.

DIGITAL TECHNOLOGIES

Digital technologies refer to a subset of electronic technologies that include hardware and software used by individuals for educational, social and/or entertainment purposes in schools and at home. These technologies include:

- interactive whiteboards (also called Smartboards)
- desktop computers
- mobile devices, for example laptops, tablets, ultramobiles, mobile phones, smartphones, PDAs, game consoles
- digital recording devices, for example cameras, flipcams, voice and video recorders
- data logging equipment and associated probes
- Web 2.0 technologies and other resources on the Internet, for example information and multimedia resources, communication and collaborative resources such as *Skype*, *Moodle*, *Edmodo*, blogs, glogs and wikis, concept-mapping tools such as *SpicyNodes*, *cMap* and *Bubble.us* and storage spaces such as *Dropbox* or *SkyDrive* and
- the variety of software packages for learning that are either commercial, downloadable for trial for fixed periods of time, or are totally freely accessible on the Web.

The use of handheld mobile devices for learning while still considered relatively new in educational institutions, is increasingly being explored for learning in the classroom. Examples include the use of PDAs (e.g. Fluck, 2008; Ng & Nicholas, 2009; Shih, Chuang & Hwang, 2011; Zurita & Nussbaum, 2004), mobile/smart phones (e.g. Chng, Wati Abas, Goolamally, Yusoff & Singh, 2011; Grant, Daanen & Rudd, 2007; Hartnell-Young & Heym, 2008; Keogh, 2011; Stewart & Hedberg, 2011) and iPods/MP3 players (e.g. Belanger, 2007; Caron, Caronia & Gagné, 2011; Crispin & Pymm, 2009; Murray, 2011). The use of mobile devices as pedagogical and social learning tools has been discussed by Nicholas and Ng (2009) and Ng, Nicholas, Loke and Torabi (2010).

The modes of digital technology usage and the common types of educational software and hardware associated with them for science learning are shown in Table 3. The tools used for science learning falls into the main categories of those (i) for capturing data, image, audio and video-recordings during practical work in the laboratory or fieldwork (ii) for data analysis (iii) for presentation (iv) that provides information for research purposes, for example the Internet and other digital references (v) for communication and collaboration and (vi) that assist with concept development and learning, for example multimedia learning objects. Examples of hardware and software associated with these categories are shown in Table 3.

Table 3. Modes of digital technology usage in science classroom

	Activity	Hardware/software
Capture tools	Data capture	dataloggers, video capture device
	Image/video capture	digital camera, video recorder, digital microscope, flipcam, scanner
	Audio capture	digital recorder, smartpens,
Data analysis		spreadsheets, statistical packages, Mathematics software packages, purpose written code – self written or provided
Productivity and presentation		word processor, *Smartboards*, presentation software e.g. *PowerPoint*, *Flash*, *Prezi*, *Movie Maker/iMovie*, *Photoshop*; web pages (e.g. free hosting sites like yolasite.com; weebly.com; wikispaces.com)
Research		World Wide Web; reference software e.g. *Grolier Multimedia* or *Britannica Encyclopedia* and dictionaries; online science encyclopedia (e.g. at http://www.encyclopedia.com/c/2977-science-and-technology.html)
Communication and collaboration		email, sms, learning management systems, chat rooms, discussion boards, conferencing systems e.g. *Elluminate* and *Wimba*; Web 2.0 technologies for collaborative work e.g. wiki for WebQuest tasks; *VoiceThread* to share science reports and assignments or science related experiences through digital storytelling; *Fakebook* for developing science historical events; *GoogleDocsApps* for creating and editing science assignments, presentations and spreadsheets online
Content learning	Concept development	*Microworlds* for development of abstract ideas through Logo programming language; *Scratch* and *Flash* for creating animations; *Inspiration, SpicyNodes* for concept mapping; *Smartboards* for interactive teaching and learning; simulation software e.g. *Sim* series such as *SimLife* and *SimAnimal*; *Crocodile* series such as *Crocodile Clips* and *Crocodile Chemistry*; other science simulation applets on the Web; quizzes e.g *Hot Potatoes* and *SurveyMonkey, Quizlet, Quia*
	Mastering content	tutorial software, drill and practice software; games; problem solving software

As the term 'information and communication technology' (ICT) is commonly used in educational settings and is usually associated with the computer and other digital devices, this book will embrace the term to mean the digital technologies that support information retrieval and creation, and communication through digital means such as email, social networking sites and learning management systems. While information could be found in paper-based books and communication could mean the use of pen, paper or the white/blackboard technology, they are not digital resources/equipment and do not fall into the category of digital technologies or ICT. For the remaining of this book, ICT and digital technologies could be used interchangeably to mean similar things, depending on the contexts.

DIGITAL LITERACY

With the rapidly changing landscape of digital technology in society, a range of terms related to its literacy has been proposed in the literature, for example ICT literacy, information technology literacy, digital literacy, technology literacy, media literacy, information literacy, net literacy, online literacy and new literacies. Attempts at distinguishing them have been made, for example, by Markauskaite (2006), Ng (2006) and Oliver and Tomie (2000). Frameworks for the different literacies have also been proposed, for example the US Educational Testing Services (ETS, 2007) identified five components for ICT literacy as essential for functioning in a knowledge society: (i) accessing (ii) managing (iii) integrating (iv) evaluating and (v) creating information. The complexity of the knowledge and expertise from (i) to (v) is incremental, similar to the levels in Bloom's taxonomy. ETS further proposed that there are two sets of foundational knowledge and skills underlying ICT literacy, these being:

- cognitive proficiency, the foundational skills of everyday living, these being literacy, numeracy, problem solving and spatial/visual literacy
- technical proficiency, the basic components of digital literacy that includes basic knowledge of hardware, software applications, networks and elements of digital technology.

While the ETS (2007) places digital literacy as a subset of ICT literacy, there are other broader definitions of digital literacy that places the ETS' five components as a subset of digital literacy. Aviram and Eshet-Alkalai (2006) and Eshet-Alkalai (2004) have reviewed the various definitions for digital literacy in the literature and concluded that within the broadness of the definitions of digital literacy, singly or in combination, it covers meanings that are technical, cognitive, psychological and/or sociological. An example of a broad, encompassing definition of digital literacy is from the British

Futurelab' handbook on *Digital Literacy Across the Curriculum* (Hague & Payton, 2010, p.2) which states that:

> Digital literacy is an important entitlement for all young people in an increasingly digital culture. It furnishes children and young people with the skills, knowledge and understanding that will help them to take a full and active part in social, cultural, economic, civic and intellectual life now and in the future...To be digitally literate is to have access to a broad range of practices and cultural resources that you are able to apply to digital tools. It is the ability to make and share meaning in different modes and formats; to create, collaborate and communicate effectively and to understand how and when digital technologies can best be used to support these processes.

Another similarly broad definition of digital literacy is one that is formulated by the *European Information Society* (Martin, 2005, p. 135) which states:

> Digital Literacy is the awareness, attitude and ability of individuals to appropriately use digital tools and facilities to identify, access, manage, integrate, evaluate, analyse and synthesize digital resources, construct new knowledge, create media expressions, and communicate with others, in the context of specific life situations, in order to enable constructive social action; and to reflect upon this process.

Both definitions bear a similarity to scientific literacy in that educating individuals to have digital literacy skills and knowledge is, among other purposes, to enable them to participate actively as citizens in society. Both emphasise the ability to create meanings and communicate effectively with others (through digital tools).

At a more specific level, Eshet-Alkalai (2004) suggested that there are five types of literacies that are incorporated within the term 'digital literacy': (i) photo-visual literacy (ii) reproduction literacy (iii) branching literacy (iv) information literacy and (v) socio-emotional literacy:

(i) *photo-visual literacy.* This is a learning-to-read from visuals and is a cognitive skill that uses "vision to think". It helps individuals to 'read' intuitively and understand messages that are presented in a visual-graphical form. People who have high photo-visual literacy usually have good visual memory and strong intuitive-associative thinking

(ii) *reproduction literacy*. This is the art of creative duplication. The areas which are largely influenced by reproduction literacy are in the arts and in academic work where there is a lot of writing. Modern digital tools have the capabilities to edit or combine/recombine new and pre-existing materials (text, audio, video, images) into new works of art or writing

(iii) *branching literacy*. One of the affordances of technology is the creation of non-linear medium of information (hypermedia) where multimedia (text, graphics, audio and video) and hypertext are intertwined. Branching literacy, also called hypermedia literacy, is the literacy of hypermedia and the thinking associated with working in a hypermedia environment. In hypermedia, the individual is able to navigate through the displayed text freely, pausing to investigate related meanings for a term/concept/image that has a hyperlink before continuing with the original text. Learning in this way is just-in-time, spontaneous and supporting coherent learning. The use of hypertexts requires skills such as the ability to remain oriented and avoid getting disoriented while navigating through the different hyperlinked pages (Eshet-Alkalai, 2004 citing Daniels, Takach & Varnhagen, 2002; Horton, 2000; Piacciano, 2001). Branching-literate individuals have good spatial orientation and the ability to create mental models, concept maps and other forms of abstract representations in hypermedial environments (Eshet-Alkalai, 2004 citing Lee & Hsu, 2002)

(iv) *information literacy*. This is the literacy associated with critical thinking and the ability to search, locate and assess Web-based information effectively. Assessing information involves the ability to critique through analysing and evaluating digital content for accuracy, neutrality, currency, reliability and level of difficulty. Information literacy is associated with the art of skeptism (Eshet-Alkalai, 2004), similar to the art of always questioning information (Aviram & Eshet-Alkalai, 2006)

(v) *socio-emotional literacy*. This is the literacy associated with the emotional and social aspects of online socialising, collaborating and doing day-to-day chores e.g. banking and purchasing online. It requires the ability to be highly critical and analytical, to avoid online 'traps', for example being able to identify pretentious people in the chat rooms and avoiding hoaxes and viruses. Socio-emotional literacy skills develop as individuals mature and require a good level of photo-visual, branching and information literacy skills, all of which can be taught.

CONCEPTUAL FRAMEWORK UNDERPINNING DIGITAL LITERACY

The framework for digital literacy for this chapter will draw together the broad definitions and the specific features of the different computer-associated literacies mentioned above. These will be discussed in light of the technical, cognitive and social/emotional dimensions of digital literacy. The multiple literacies associated with digital literacy, such as those proposed by Eshet-Alkalai (2004), will be incorporated into the framework.

As discussed in Chapter 2, a substantial part of an individual's learning is done informally outside of school. As the role of mobile devices (particularly mobile phones and smartphones) increasingly influences this learning, the framework for digital literacy in this chapter will also include mobile learning literacy or 'mlearning literacy' - the digital literacy associated with socialising and learning with mobile devices. While many of the digital literacy skills that are related to desktop/laptop-based learning are transferable to mlearning literacy, there are specific skills that are mobile-based, for example, text messaging. There are also limitations that are imposed by the small screen size of mobile devices and their less powerful processing ability that need to be considered. Tables 4-6 lists the digital literacy characteristics related to desktop/laptop learning and mlearning and show their similarities and differences in the three dimensions of (i) *technical*: technical and operational skills, shown in Table 4 (ii) *cognitive*: critical thinking and evaluative skills, shown in Table 5 and (iii) *social-emotional*: social and cybersafety skills, shown in Table 6. Each of these dimensions will be discussed in more detail in their respective sections and across both desktop/laptop and mobile learning.

Generic skills that are identified as applicable to both desktop/laptop learning and mlearning literacies include:

- knowing about the machine that the learner is using
- developing proficient technical skills
- selecting the right tool/applications to explicit represent the learner's thinking and understanding, including newly constructed knowledge

- critically analysing data received (both information and conversational data) and
- exercising caution and appropriate behaviour while socialising in online communities.

The mobile user however needs to be able to differentiate between the different devices and how they differ technically and functionally as well as understanding the affordances and limitations of similar software in mobile devices, for example *Excel* vs *mobile Excel.*

1. Technical Dimension of Digital Literacy: Technical and Functional Skills

The technical dimension of digital literacy is shown in Table 4. At the most basic level, a digitally literate person should be able to connect together a functional computer system for his/her own personal use, for example a desktop to a printer. The ability to read manuals to conduct basic technical activities is part of being digitally literate. In addition to or instead of reading manuals, digitally literate individuals are able to search for online resources that could assist with troubleshooting. Keying the right questions or phrases into a search engine would enable the individual to retrieve responses in the form of text, images and/or videos that will provide explanations to address the problem. An understanding of the use of and the regular updating of anti-virus software to avoid spam and viruses is also part of being digitally literate technically.

Another aspect of technical digital literacy is the ability to use application software, tools that enable the individual to perform and accomplish specific tasks. The common types of educational software that are used for science learning are:

i. word processing software such as *Microsoft Word* within which there are various software applications that enhance science learning, for example *SmartArt* with templates for drawing Venn diagrams, flowcharts, pyramids etc. to show different types of relationships. Being digitally literate means being aware that there are other tools that are able to do similar functions. For example the drawing of a Venn diagram could be done using the template found in *Word*'s *SmartArt* 'relationships' section. Depending on the purpose of the diagram, they could be constructed using other software such as *Microsoft Publisher* or *PowerPoint*. The choice of software depends on the familiarity of the individual with available tools and the complexity of the task. As an example, Figure 1 shows a diagram of a 'Relationship framework between stakeholders in implementing mobile learning in schools' that is constructed in *PowerPoint*. It includes Venn diagrams, block arrows, text and other features. Given the available tools of *Word*, *PowerPoint* and *Publisher*, I have found that it is easier to construct this quite complex relationship framework in *PowerPoint* rather than starting with the Venn diagram template provided by *Word* or using *Microsoft Publisher*. The completed diagram in a *PowerPoint* slide could be easily copied and pasted onto a *Word* document as a single image. As an image, the diagram can then be easily moved around to different locations of the *Word* document.

Table 4. The technical dimension of digital literacy associated with desktop/laptop and mobile learning

Skills & knowledge	Literacy	Desktop/Laptop learning literacy	mLearning literacy
TECHNICAL: TECHNICAL AND FUNCTIONAL SKILLS	Able to connect and use input and peripheral devices	for example mouse, QWERTY keyboard, earphones/headset, VGA connector, printer, USB drive, external speakers, smartboard	such as touch screen keypads, navipad, stylus, half-QWERTY keyboard, earphones, Bluetooth headset, USB connector
	Knowledge of working parts	in particular the common ones in 'Control Panel' for example: wireless, sound, display; use of multiple windows for multitasking	for example, infra-red, Bluetooth, wireless, memory card, data synchronization, use 'Settings' to control features such as sound, brightness; use of multiple windows. General understanding of standard device keys and controls such as soft-keys/navipad, how they differ for various devices and ability to successfully locate the equivalent keys on new devices to make calls or send SMS
	Able to troubleshoot by	reading manuals, accessing local Help functions and/or web-based resources (e.g. YouTube) for assistance	reading manuals, accessing local Help functions and/or web-based resources (e.g. YouTube) for assistance

CHNICAL: TECHNICAL AND FUNCTIONAL SKILLS	Able to operate adequately:	understanding file structures e.g. folders and directories and managing where files are located in the drive, managing data transfer that includes understanding file sizes e.g. audio/video files often in MB and space to hold files; finding, downloading and installing applications (and uninstall when not needed), update /change user account information on the Internet; understand data charge costs associated with downloading data; creating shortcuts, embedding links, sending and retrieving attachments via email and/or Dropbox, opening them with appropriate applications, unzipping folder; creating a CD; for computers/laptops with infra-red and Bluetooth, use them for file transfer; cut, copy, paste, save and knowing where to find them later; knowing about and able to locate frequently used user interface elements i.e. cues that define interactivity e.g. menu, sizing, dragging, scrolling, using sliders, collapsible lists; understanding tabs and their relationship to content; use of multiple windows for multitasking set up and use communication/social networking tools e.g. emails, web mail, VOIP (e.g. Skype), blogs, wikis, Facebook etc; ensuring that anti-virus software is regularly updated to avoid spam and viruses.	understanding file structures e.g. folders and directories and managing where files are located in the drive/memory card and how to access your drive via your computer (understanding about operating system compatibility), managing data transfer that includes understanding file sizes e.g. audio/video files often in MB and space to hold files; finding, downloading and installing applications (and uninstall or delete), update /change user account information on the device and on the Internet; understand data charge costs associated with downloading data; sending and retrieving attachments via email and/or Dropbox, opening them with appropriate applications, unzipping (e.g. using Pocket RAR), use infra-red and Bluetooth for file transfer; understand limitations of word processing, spreadsheet and presentation tools and compatibilities of operating systems and transferability of files between systems; knowing about and able to locate available user interface elements i.e. cues that define interactivity e.g. menu, sizing, scrolling, using sliders, understanding tabs and their relationship to content; use of multiple windows for multitasking set up and use communication tools e,g. emails, web mail, VOIP, blogs, wikis, Facebook, Twitter, send SMS, MMS; ensuring that anti-virus software is regularly updated to avoid spam and viruses.

ii. spreadsheets e.g. *Excel* and *Numbers* for recording, analysing and graphing science experimental data; analysis software packages, for example *Statistical Packages for the Social Sciences* (*SPSS*) for quantitative data analysis and *NVivo* for qualitative data analysis. The latter two can be used for research purposes on a larger scale where more complex analysis of data is required.

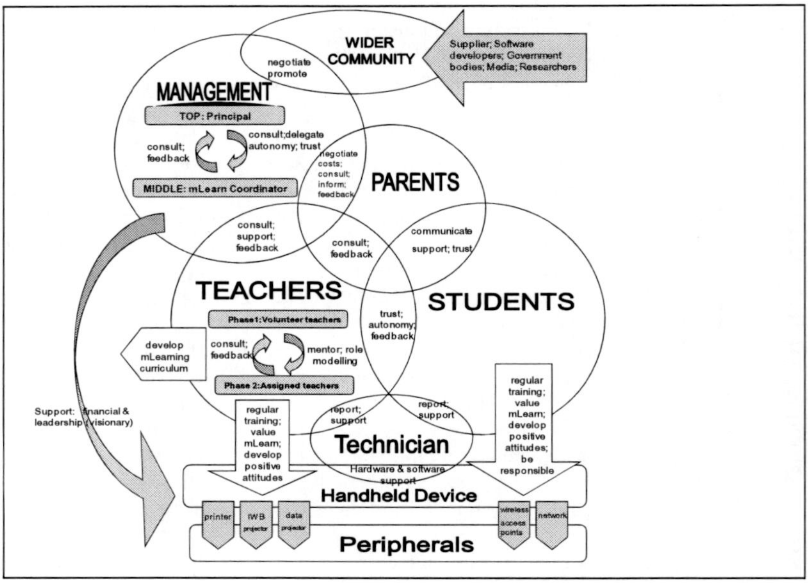

Figure 1. An example of a framework constructed on a *PowerPoint* slide and pasted onto a *Word* document. This framework shows the relationships between stakeholders in implementing mobile learning in schools.

iii. software for creating databases e.g. *FileMaker Pro* and *Microsoft Access* for data storage. However less complex school-based science databases could be created in *Excel*, for example a glossary for key words studied in each topic or the classification of animal and plant species.

iv. concept mapping software e.g. *Inspiration*, *Kidspiration*, *FreeMind* and online software such as *cMap* and *SpicyNodes*. Concept maps are useful for showing the connections between key science terms and concepts, for example demonstrating understanding between photosynthesis and respiration or for forces and motion.

v. software e.g. *SurveyMonkey*, *Hot Potatoes, Quizlet, QuizCreator* and *Quia* for creating worksheets such as crossword puzzles, study games or tests in multiple choice and short answer formats.

vi. drill and practice, tutorial and problem-solving software e.g. Chemistry tutorial at http://science.widener.edu/svb/tutorial/ or *Edheads Simple Machine* game at http://www.edheads.org/activities/odd_machine/index.htm.

vii. presentation and associated editing and creative software e.g. *PowerPoint*, *Prezi*, *MovieMaker*, *iMovie,PhotoStory* and *PhotoShop.*

viii. referencing software e.g. *Grolier Multimedia* or *Britannica Encyclopaedia* and other online science dictionaries e.g. http://www.thesciencedictionary.com/ and science/engineering encyclopaedia e.g. http://www.diracdelta.co.uk/.

ix. interactive software such as science simulation, animations and games. An example of a science simulation is the mathematics and physics shareware at http://scienceshareware.com/indexSub.htm, where interactive simulations on a number of mathematics and physics topics are provided cost-free. Figure 2(a) shows a screenshot of a simulation from this website on how a light pulse while travelling through a fibre optics cable. An example of a science animated site where the resources are free is at http://www.kscience.co.uk/animations/anim_1.htm. The screenshot shown in Figure 2 (b) shows an interactive animation of parts of the kidney and a nephron where rolling the mouse over a structure will show the name of the structure and clicking on the structure will show an animation of how it works. An example of free science games for junior students can be found at http://www.sciencekids.co.nz/gamesactivities.html.

x. online collaborative tools such as *Popplet*, *Fakebook*, *VoiceThread* and those for creating wikis, blogs and glogs (online interactive posters).

There are both technical and cognitive skills involved in using the software packages described above effectively. Technical skills involve installing programs or downloading them from the Web, extracting (if zipped) and installing them onto the computer. Other technical skills involve operating the software through understanding the features of the software and being able to navigate around, for example when to click or double click on a function, or use highlight, drag, expand or drop features. There are specific functions for each type of software that the individual needs to learn and the more practice an individual has with the software, the better skilled (s)he will be. With the availability of help within the software packages and on the Web, digitally literate individuals should be able to find information that will assist them to learn how to operate specific software. For example, an individual could literally learn all about *Excel* from *YouTube* videos by him/herself. By typing 'using microsoft excel' into *Google* and selecting 'videos', more than 6,000 results show up. There are two sets of tutorials that systematically take the learner through the functions of *Excel*. One set, produced by *Motion Training*, have 19 tutorials called 'Microsoft Excel Tutorials for Beginners' plus another additional three tutorials for 'Excel VLOOKUP'. Similarly, *ExcellsFun* has produced a series of 23 tutorials called 'Excel Basics'. These two sets of tutorials, singly or in combination, would provide the necessary information for beginners who wish to learn about the use of spreadsheets. By going through two sets of tutorial, the individual would be able to look at different examples and different styles of presentation to effectively learn about the functions of *Excel*. Part of being digitally literate is the ability to learn independently with digital resources that are available both online and offline. However, using these resources require cognitive skills to learn intelligently and selectively. The individual needs to be aware of the authority and accuracy of the videos that they are viewing. Hence looking up the background of the person or institution that uploaded the videos would provide some information of the authenticity of the videos. To ensure the accuracy of the information provided, viewing a few similar videos on the same topic and comparing the information between them is part of what a digitally literate person would do. This is an aspect of the cognitive dimension of digital literacy which is discussed in the next section.

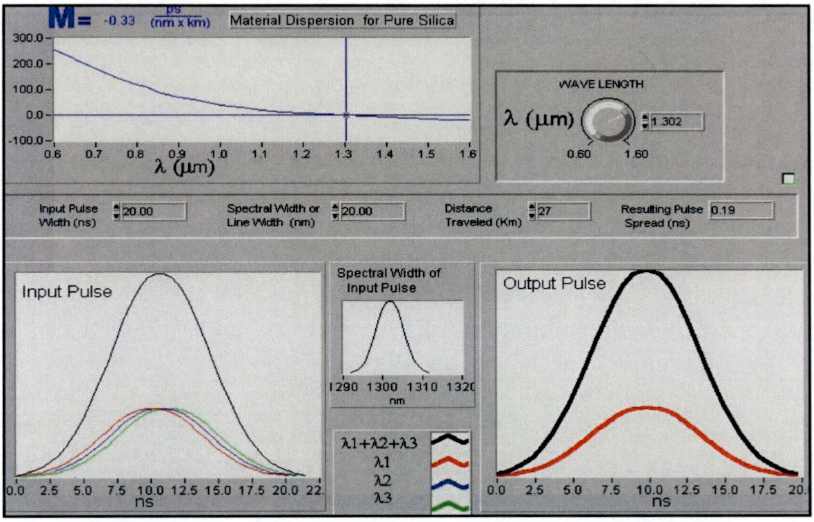

Figure 2a. Simulation of how a light pulse while traveling through a fiber optics cable. This freeware is sourced at http://scienceshareware.com/dispersion_OV.htm.

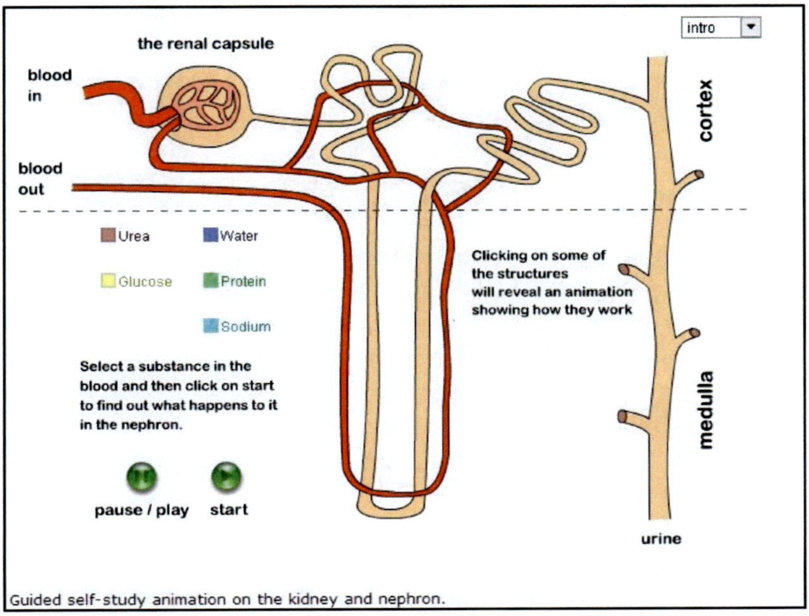

Figure 2b. Interactive animation of a nephron. This freeware is sourced at http://www.kscience.co.uk/animations/anim_1.htm.

2. Cognitive Dimension of Digital Literacy: Critical Thinking and Evaluative Skills

The cognitive dimension of digital literacy is shown in Table 5. One of the cognitive skills that is associated with using software is the ability to evaluate and select appropriate programs to learn with. Within each software program, the digitally literate individual will be able to select the most appropriate feature/functions to solve a problem or to demonstrate understanding of knowledge acquired. This selective process is vital as the number of applications that could do similar tasks seems to be increasing. For example, under what circumstances would an individual use *OneNote*, *VoiceThread*, *Moodle*, *Elluminate*, *Wimba*, *Skype*, *Ning*, *Popplet,* blogs, wikis, *Tweeter*, *Facebook* or *GoogleApps*? All these technologies enable collaboration between learners. Some have trial periods or are free with limited features until you pay for the full versions, some of these programs allow for synchronous communication while others are only asynchronous. Some of these programs lend themselves better for socialising and networking but could equally be adapted for learning – this will depend on how they are used. Many of these programs have multimedia capabilities where text could be reinforced with images and audio or video features. Selecting the most appropriate software to do a given task would require an understanding of the capabilities of the available tools and how they could be used to deliver the outcomes required. A good way to start would be to read about these tools on the Internet and ask targeted questions that will provide information on which program would fulfil the desired objectives.

A crucial set of cognitive skills necessary for the development of digital literacy are those associated with using information from the Web. Much has been written about these skills that are needed to search, assess and use web-based information and resources (for example, McKenzie, 1998; Ng, 2006; Ng & Gunstone, 2003; Wallace, Kupperman, Krajcik & Soloway, 2000). These and other skills and knowledge that are associated with using web-based resources, such as knowledge of ethical, moral and copyright issues, are elaborated below.

Search and Assess Information on the Web

The vast amount of resources on the Web that individuals can access at no cost in their own time and place provide them with opportunities to work with knowledge that is ever-changing and pluralistic (Angelique and Lim, 2011). One of the advantages that web-based resources offer is the presentation of materials in varied formats for learning. The formats include text-based written materials, visual-based (for example, an image of the digestive system), audio-based (for example, a podcast on oxidation/reduction) and multimedia-based that enable abstract concepts to be made more concrete (for example, simulations of protein synthesis or heat transfer). Another advantage is that web-based learning environments are able to support communicative activities such as emailing, chat rooms and blogs on social networking sites or learning management systems for the sharing of ideas across distance easier and quicker. These features provide opportunities for flexible learning for individuals to carry out independent research and learning whenever and wherever they choose to, including through the use of mobile phones and handheld computers. The variety of materials also means that a commensurate sophistication is required in the accessing, comparing and evaluating the content of the resources.

Digital literacy skills that are necessary for researching and using web-based resources include being able to:

(i) handle the Web appropriately, such as being able to distinguish between the different search engines and using the more 'suitable' search engine for a particular purpose
(ii) narrow down the search using appropriate keywords to maximise precision and to reduce the number of pages that the learner has to read, for example, using multiple (3-4) keywords in the search would yield better results than a single keyword
(iii) critique the content of webpages in terms of accuracy, currency, reliability and the level of difficulty. Studies by Wallace, Kupperman, Krajcik and Soloway (2000) and Hoffman, Wu, Krajcik, and Soloway (2003) have shown that the use of 'search and assess' inquiry skills where information at one site is critically analysed before doing another search, is necessary for effective understanding of the content from web-based resources

Table 5. The cognitive dimension of digital literacy associated with desktop/laptop and mobile learning

Skills & knowledge	Literacy	Desktop/Laptop learning literacy	mLearning literacy
COGNITIVE: CRITICAL THINKING & EVALUATIVE SKILLS	Use the Internet effectively for information gathering and synthesis by	being able to search, locate and assess web-based information i.e. use appropriate browser and search engine, critique information through analysing and evaluating the digital content for accuracy, currency, reliability and level of difficulty; being able to navigate through hypermedia environments to construct knowledge; understanding that people behind the scene writing the information have their own motivations and being ablt to critically evaluate whose voice is heard and whose is not is important for learning as neutrally as possible; knowing about the ethical and moral issues associated with writing that uses web-based resources, for example copyrights and plagiarism; synthesising new understandings using appropriate online (e.g. wiki) or offline (e.g. Word, PowerPoint) tools that will convey the meanings in the best manner; understand the terms and conditions well so that legal liability is avoided, ensuring safety e.g. when meeting the purchaser or seller either online or offline to exchange goods and payment e.g. *eBay*	being able to search, locate and assess web-based information i.e. use appropriate browser and search engine, critique information through analysing and evaluating the digital content for accuracy, currency, reliability and level of difficulty; understanding that most websites are not mobile friendly and created for mobile devices and a fair amount of scrolling (up and down, left and right) will be needed to read information on the web; understanding that mobile web pages could contain reduced versions of non-mobile web pages and that the balanced perspective of the article may not be complete; understanding that people behind the scene writing the information have their own motivations and being able to critically evaluate whose voice is heard and whose is not is important for learning as neutrally as possible; knowing about the ethical and moral issues associated with writing that uses web-based resources, for example copyrights and plagiarism; synthesising new understandings using appropriate online (e.g. wiki) or offline (e.g. Word, PowerPoint) tools that will convey the meanings in the best manner and understand that the capacity of productivity application are reduced in mobile devices; understand the terms and conditions well so that legal liability is avoided, ensuring safety e.g. when meeting the purchaser or seller either online or via mobile phone to exchange goods and payment e.g. *eBay*
	Be multiliterate in being able to	decode information that are text-based and information from images, sound bytes (e.g. podcasts), videos, maps and models – these involve multiliteracies skills that are linguistic, visual, audio, spatial, gestural (as captured in videos) and multimodal (as in multimedia resources)	decode information that are text-based and information from images, sound bytes (e.g. podcasts), videos, maps and models – these involve multiliteracies skills that are linguistic, visual, audio, spatial, gestural (as captured in videos) and multimodal (as in multimedia resources)

(iv) use images (*Google Image*), maps (*Google Maps*), videos (*YouTube*) and/or podcasts (for example, *iTuneU*) to help with understanding text-based information. The learning from texts, images, maps, videos and podcasts involve linguistic, visual, spatial, audio and multimodal literacies – these are collectively a part of the framework known as 'multiliteracies' (New London Group 1996) and will be discussed in the next chapter.

(v) The enormous number of videos on *YouTube* means that, as noted in the previous section, skills to search and assess the authority and authenticity of the videos need to be developed. There are, however, other websites that are dedicated to educational videos only, such as the *Khan Academy* at www.khanacademy.org/. The *Khan Academy* is a not-for-profit organization with the aim of making education better by providing free resources to anyone. The organisation has over 2200 educational videos (TED 2011[2]) with about a million students using it every month. The videos, stored on *YouTube* teach a range of concepts in the areas of Mathematics, Astronomy, Biology, Chemistry, Physics, Economics, History, Finance and Computer Science. Accessing *YouTube* videos via educational sites like the *Khan Academy* is a safer way of searching and learning from videos as opposed to random searches on *YouTube*

(vi) understand ethical and moral issues associated with using web-based resources, for example copyrights and plagiarism and

(vii) select appropriate online tools to create content that will convey the meanings as intended.

These are capabilities akin to Eshet-Alkalai's (2004) information literacy and branching literacy in the 'search and locate' aspect of finding appropriate information on the Web. These capabilities are also echoed by the UK Society of College, National and University Libraries (SCONUL) who has produced their 'seven pillars' model for information literacy as (Town, 2000 cited in Martin & Grudziecki, 2006):

1. Recognising an information need;
2. Identifying what information will fulfil the need;
3. Constructing strategies for locating information;
4. Locating and accessing the information sought;
5. Comparing and evaluating information obtained from different sources;
6. Organising, applying and communicating information;

[2] http://www.youtube.com/watch?v=gM95HHI4gLk&feature=player_embedded#at=93

7. Synthesising and building upon information.

These are higher order thinking skills that are described by McKenzie (1998) as ‘infotective’ skills that are necessary for all students to develop. McKenzie described today’s students as ‘free-range students’ who have unlimited opportunities to graze the Net freely and that as ‘infotectives’, they should be able to direct their own learning by applying skills to search and sort, to evaluate their way through the piles of often fragmented information and to rearrange them in some meaningful format. In his article *Grazing the Net:Raising a Generation of Free Range Students* available at http://www.fno.org/text/grazing.html, McKenzie asserted that:

> We must also give students the tools to overcome the weaknesses of the new information sources...The extensive information resources to be found in cyberspace are both a blessing and a curse. Unless students possess a toolkit of thinking and problem-solving skills to manage the inadequacies of the information landfills, yard sales, gift shoppes and repositories so prevalent on the "free Internet," they may emerge from their shopping expeditions and research efforts bloated with techno-garbage, information and junk food or info-far.

One of the tools that students need to be equipped with in carrying out information literacy effectively is critical literacy, a concept which is discussed in a later section of this chapter. McKenzie further argued that the decision making skills in carrying out infotective actions are the same sorts of skills that students will use in deciding about important issues that affect their own lives. He added that to be successful with the promoting of the digital literacy skills mentioned above, there need to be an emphasis on the development of questioning skills which is an aspect of critical literacy.

3. Social-Emotional Dimension of Digital Literacy: Social and Cybersafety Skills

The social-emotional dimension of digital literacy is shown in Table 6. On a daily basis, millions of young people ‘meet’ online to chat, exchange ideas, communicate socially and collaborate on projects. Web 2.0 technologies such as sites at *Wikispaces*, *Flickr*, *MySpace*, *Google+*, *Blogspot*, *Facebook* and *YouTube* enable individuals to contribute to networked communities for learning and/or for socialising. Web 2.0 is becoming a lifestyle for young people who are accessing the Web to send emails, seek information, purchase science-related and other goods, do online banking, chat online, post questions, contribute solutions, post photos and videos, download games, videos and music and write blogs and wikis to share ideas and opinions. In science learning, the interactivity of Web 2.0 technologies enable collaborative learning in online communities, for example joining blogs that discuss science topics of interest or a science-related contemporary issue such as carbon taxing, contributing to wikis of a scientific nature or collaborating with scientists on specific projects.

As Web 2.0 environments become more and more of a lifestyle for young people, the issue of cybersafety and potential risks in participating online is increasing (Conroy, 2007; Hanewald, 2008). Cybersafety is about keeping safe online. The potential risks that young people face online include being bullied, stalked, harassed and exposed to identity fraud and inappropriate materials such as pornographic, illegal materials, spam and computer viruses. Hanewald (2008), in her literature review on the research into cyber safety cited work conducted in the US (Finkelhor, Mitchell & Wolak, 2000; Wolak, Mitchell & Finkelhor, 2006) and in the UK (Smith, Mahdavi, Carvalho & Tippett, 2006) to show an increase in primary school-aged children and adolescents' experiences with cyberbullying and harassment. For example, the US studies indicated a rise in cyber bullying from 28 % in 2000 to 48 % in 2006. The various studies also indicated that text messaging, emails and chatroom postings are the most common means of cyberbullying. Unlike physical bullying where the perpetrator is known, the cyber perpetrator could be difficult to identify. The resulting impact of cyber bullying and harassment on children and young adolescents could lead to both short and long-term psychological harm which is manifested in depression and anti-social behaviour. This in turn would have negative effects on family lives and schooling.

Inculcating privacy and cybersafety literacies as well as online etiquette in students are imperative for successful and safe participation in online communities. As listed in Table 6, a digitally literate individual will observe 'netiquette' by applying similar rules to face-to-face communication such as respect and using appropriate language and words in order to avoid misinterpretation and misunderstanding. Being digitally literate means having an awareness of cybersafety (Becta, 2006) and the need to protect his/her own safety and privacy by keeping personal information as private as possible.

Table 6. The social-emotional dimension of digital literacy associated with desktop/laptop and mobile learning

Skills & knowledge	Literacy	Desktop/Laptop learning literacy	mLearning literacy
SOCIAL-EMOTIONAL: ONLINE SOCIAL SKILLS & SAFETY	Use the Internet responsibly for communicating, socialising and learning by	observing 'netiquette' by applying the similar rules as in face-to-face; communication such as respect and using appropriate language and words to avoid misinterpretation and misunderstanding; balancing the amount of time spent on social networking sites in managing costs, time to do other things and prevention of social networking addiction; protecting own's safety and privacy by keeping personal information as private as possible and not disclosing any more personal information than is necessary; interpreting social network messages in terms of the tone of the message (e.g. use of bold letters and symbols) and underlying meanings correctly; recognising when (s)he is being threatened and know how to deal with it, for example whether to ignore, report or respond to the threat	observing 'netiquette' by applying the similar rules as in face-to-face; communication such as respect and using appropriate SMS language to avoid misinterpretation and misunderstanding; being aware that SMS language can impact formal language usage in classroom situations; balancing the amount of time spent on social networking sites and the amount of text messages sent in managing costs, time to do other things and prevention of social networking addiction; protecting own's safety and privacy by keeping personal information as private as possible and not disclosing any more personal information than is necessary; interpreting SMS and social network messages in terms of the tone of the message (e.g. use of bold letters and symbols) and underlying meanings correctly; recognising when (s)he is being threatened and know how to deal with it, for example whether to ignore, report or respond to the threat.

While the virtual world provides young people with opportunities to try out new ideas and take on different personas, a digitally literate individual should be aware that there could be possible undesirable consequences for the actions that they take. (S)he should appreciate the fact that his/her online conversations are recorded permanently and what (s)he may think is private is in fact public. A digitally literate individual should be able to recognise when (s)he is being threatened and know how to deal with it, for example whether to ignore, report or respond to the threat. The fostering of respect and responsibility when communicating through digital technologies and the understanding of ethical and cultural issues associated with digitally-based environments should be a part of the digital literacy education of students (Berson & Berson, 2003).

In summarising the above discussions, being digitally literate requires the development of a set of key skills that are technical, cognitive and social-emotional. The basic skills that a digitally literate person should be able to demonstrate are:

(i) carry out basic computer-based operations and access resources for everyday use
(ii) search, locate and assess information effectively for both purposes of research and content learning where assessing information involves the ability to critique through the analysis and evaluation of digital content for authenticity
(iii) select and develop competency in using the most appropriate technological tool or features to complete a task, solve problems or create products that best demonstrate new understandings
(iv) behave appropriately and communicate effectively in social network communities and protect oneself from harm in digitally enhanced environments.

Teaching these skills in context and providing opportunities to practise them in ways that demonstrate their significance in the choices made and in using them appropriately are essential and invaluable to both the personal and academic development of students.

THE MULTIPLE LITERACY DIMENSIONS OF DIGITAL LITERACY

The dimensions of digital literacy and the associated multiple literacies, as discussed in the preceding paragraphs, are summarized in Figure 7. The three dimensions of digital literacy as shown in the three circles are the cognitive, technical and social-emotional dimensions. Within each of these dimensions are the supporting literacies. For the cognitive dimension, one of the supporting literacies is information literacy – the ability to seek out information and critically analyse and use them to achieve a purpose. This ability is further supported by multiliteracies – the ability to decode meanings from textual, visual, audio, spatial and gestural forms of representations. The other two circles represent the technical and social-emotional dimensions of digital literacy. The former is associated with operational abilities to use technology in a functional manner so that learning could take place. Developing technical skills for the different software would require opportunities to identify and practise useful features so that these skills become a routine part of the usage and little effort will be needed to use them during the learning. Being digitally literate technically means being aware that many skills are transferable from one application to another or from one platform to another. There will be features specific to new packages and some adaptations may be required. Within the social-emotional perspective of digital literacy is the awareness of one's own self in digitally enhanced environment, particularly in online communities where the individual needs to self-manage his/her emotions and impulses, motivation and perseverance, and to be reflective in how (s)he relates to others online socially or in collaborative team work.

The intersecting area between the cognitive and social-emotional perspectives of digital literacy involves online etiquette and cybersafety literacies where there is demonstrated ability to assess situations to ensure safe and friendly interactions such as when to reply (or not to) to an input from a stranger and under what circumstances. This would involve critical literacy and picking up cues from the content and tone of messages, including text abbreviations. Other skills include being sensitive to others' emotional state when editing and/or commenting on their work, for example in blogs and wikis.

The area overlapping the social-emotional and technical dimensions of digital literacy involves the ability to navigate through social media sites effectively and to use the technologies of these services sensibly for social interactions. Kaplan and Haenlein (2010) have identified six different types of social media:

1. for collaborative projects (e.g. *Wikipedia*),
2. blogs and microblogs (e.g. *Twitter*)
3. content communities (e.g. *Youtube*, *Ask a Scientist*[3])
4. social networking sites (e.g. *Facebook*)
5. virtual game worlds (e.g. *World of Warcraft*[4]; *Game for Science*[5]) and
6. virtual social worlds (e.g. *Second Life*).

[3] Howard Hughes Medical Insitutie at http://www.hhmi.org/askascientist/ or MadSci Network at http://www.madsci.org/

[4] http://us.battle.net/wow/en/

[5] http://www.gameforscience.ca/

Each of these types of social media could be potential science learning spaces through enquiry in the different communities (for example blogs) or through creating content/avatar in virtual spaces and communities (e.g. wikis and *Second Life*). To successfully use these technologies, it is necessary to have an understanding of each platform and how the different technical features and rules in each community work to bring about the desired social learning outcomes. For example, a general rule in 'ask a scientist' sites would require that a search through their databases first to see if the question in mind has been asked before being directing the question in the 'submit question' section. Search skills, as discussed earlier on in this chapter, would be required in thinking about the appropriate keywords or questions into the 'search' box to maximise arriving at a solution. As searching and assessing information on the Internet require technical skills such as being able to navigate systematically across Web pages and interacting with the information found, the overlapping area between the cognitive and technical dimensions would include reproduction and branching literacies. Another example of the necessary interplay between the cognitive and technical dimensions of digital literacy is in the selection of software packages to adopt to meet a particular purpose. This has been discussed in the cognitive dimension section above.

Digital literacy, as shown in the conceptual framework in Figure 3, is where the three dimensions of cognitive, technical and social-emotional intersect. Within each of these dimensions are multiple literacies supporting the respective dimensions, most of which have been discussed above. As indicated in the figure, critical literacy is central to the development of all the three dimensions of digital literacy. Its definition and implications for digital literacy are discussed in the next section.

Critical Literacy

A view of critical literacy that is appropriate for science education is summarised by the Tasmanian Department of Education (2009) in Australia. It defines critical literacy as the ability to analyse and critique the relationships between texts, language, power, social groups and social practices.

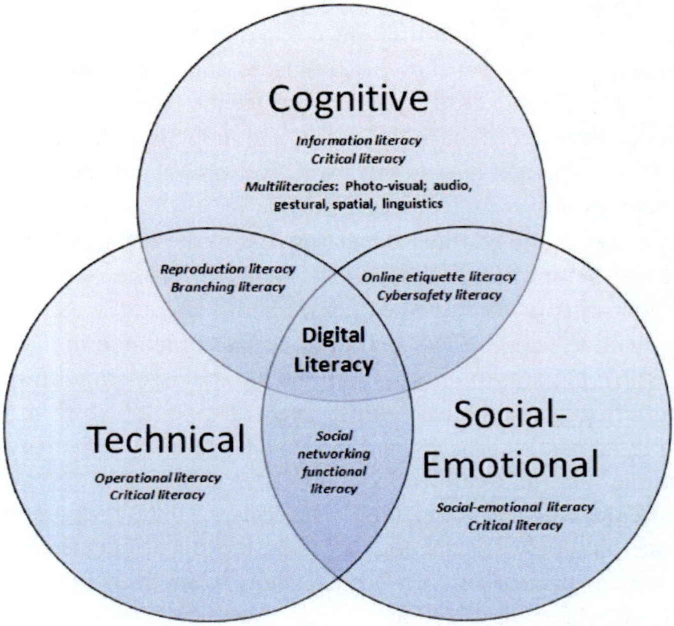

Figure 3. The dimensions of digital literacy.

It involves "ways of looking at written, visual, spoken, multimedia and performance texts to question and challenge the attitudes, values and beliefs that lie beneath the surface" (Tasmanian Department of Education, 2009, website). It is a way of understanding the relationships between power, equity and justice in society. As the amount of digital information that is available to us, through the Internet or other media outlets such as television and radio is massive, developing skills to critically analyse the multitude of 'texts' to make meaning from it is an important part of education. 'Texts' in this context includes images, music, songs, novels, conversations, movies and other multimedia materials (Coffey, nd).

Luke and Freebody (1990) developed a repertoire of capabilities in their four-part resource model for critical literacy. These are: code breaker (coding competence), meaning maker (semantic competence), text user (pragmatic competence) and text critic (critical competence). In taking the theory to the level of practise, Luke (2000, p. 454) identified the kinds of questions to ask for each component of the model as:

- *Coding practices: developing resources as a code breaker*. How do I crack this text? What are its patterns and conventions? How do the sounds and marks relate, singly and in combination?
- *Text-meaning practices: developing resources as a text participant*. How do the ideas presented in the text string together? What cultural resources can be brought to bear on the text? What are the cultural meanings and possible readings that can be constructed for this text?
- *Pragmatic practices: developing resources as a text user*. How do the uses of this text shape its composition? What do I do with this text, here and now? What will others do with it? What are my options and alternatives?
- *Critical practices: developing resources as text analyst and critic*. What kind of person, with what interests and values, could both read and write this naively and unproblematically? What is this text trying to do to me? In whose interests? Which positions, voices, and interests are at play? Which are silent or absent?

It is not within the scope of this book to discuss critical literacy at length but the practices listed above are what digitally literate individuals need to consider as they handle online science information and resources, especially those conversations and opinions on the Internet about contemporary issues such as global warming, genetically modified food and nanotechnology. 'Factual science' information would also require similar critical scrutiny, for example the work written by an academic at a university may have a different purpose and value from that written by an organization or an individual on a blog. As part of critical literacy, a search for some information on the author(s) is necessary in order to gauge from his/her background and work, the position (s)he holds in relation to the texts expressed in the resource.

Critical literacy at the technical level requires the individual to be familiar with the operational and cultural dimensions of new technologies and their associated social practices and literacies (Lankshear and Knobel, 1998). Lankshear and Knobel (1998, online) stated that:

> …we cannot produce critical readings and (re)writings of specific texts without the necessary operational capacities for accessing those texts and for framing and communicating our critical response. With conventional printed texts/typographic signs this presupposes at least requisite encoding and decoding skills. With digital texts it will presuppose also

access to a range of operational capacities with relevant hardware and software applications and procedures.

In other words, there is a need to know how each type of technology provides the capacity to shape the texts and what the technological limitations are in presenting the texts. Texts on web pages that include words, images, music etc bring a series of understanding of what a web page can do which is partly shaped by the previous social and digital experiences (including technical abilities) of the creator. A web page is also a series of social and digital interactions. It presents information in ways that are shaped by digital worlds both in terms of the content the page contains and the way the page present the information.

In drawing together the brief discussion above, the skills and abilities that students should focus on in developing critical literacy are those made explicitly by the Tasmanian Department of Education (2009). These are:

- examining meaning within texts
- considering the purpose for the text and the composer's motives
- understanding that texts are not neutral, that they represent particular views, silence other points of view and influence people's ideas
- questioning and challenging the ways in which texts have been constructed
- analysing the power of language in contemporary society
- emphasising multiple readings of texts, including considering that because people interpret texts in the light of their own beliefs and values, texts will have different meanings to different people
- having students take a stance on issues
- providing students with opportunities to consider and clarify their own attitudes and values
- providing students with opportunities to take social action

DEVELOPING DIGITAL LITERACY

The complexity of digital literacy with its various multiple literacies means that its development is an ongoing process for students. Martin and Grudziecki (2006) proposed that there are three levels of development of digital literacy (see Figure 4).

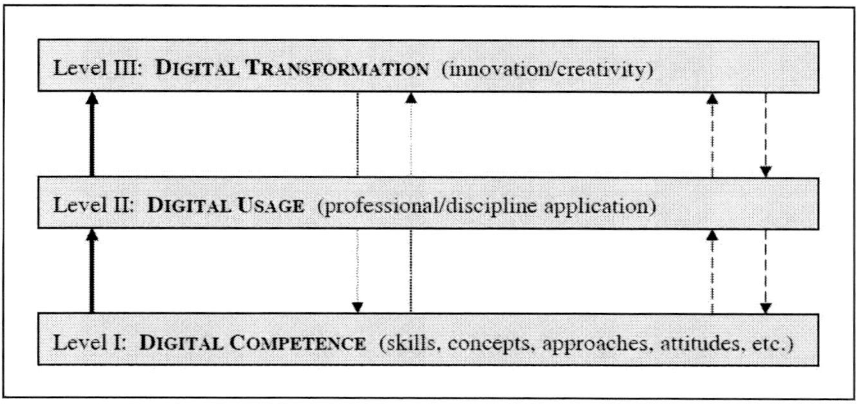

Figure 4. Levels of Digital Literacy Development (Martin & Grudziecki, 2006, p. 255).

The first level, the *Digital Competence* level, is the foundational part of the development and covers 13 processes (see Table 7, columns 1 and 2), with the 2nd to the 12th processes associated with digital literacy. These processes range from developing manual skills and skills of basic visual recognition to developing attitudes, awareness and the more critical, evaluative and conceptual skills. It is at this first level that the gradual development of the technical, cognitive and social-emotional dimensions of digital literacy and their intersecting literacies take place. Table 7 shows in the third column the mapping of the multiple literacies of Figure 3 into the processes of the Digital Literacy Development model. It shows that critical literacy plays an important role in the processes of the model. Martin and Grudziecki (2006) considered these processes as 'more-or-less sequential'. Skills and cognitions developed for specific uses are taken to the next level, the level of *Digital Usage*.

Table 7. Mapping of multiple literacies into Martin & Grudziecki's (2006, p. 257) processes of digital literacy

Martin & Grudziecki's (2006) Level 1 Digital Competency Processes		**Aspect of Digital Literacy Developed**
Process	**Descriptor**	
Statement	To state clearly the problem to be solved or task to be achieved and the actions likely to be required	
Identification	To identify the digital resources required to solve a problem or achieve successful completion of a task	Critical literacy

Martin & Grudziecki's (2006) Level 1 Digital Competency Processes		Aspect of Digital Literacy Developed
Process	**Descriptor**	**Process**
Accession	To locate and obtain the required digital resources	Critical literacy Branching literacy Operational literacy Information literacy
Evaluation	To assess the objectivity, accuracy and reliability of digital resources and their relevance to the problem or task	Critical literacy
Interpretation	To understand the meaning conveyed by a digital resource	Critical literacy Multiliteracies (visual, spatial, audio, linguistic, multimodal)
Organisation	To organise and set out digital resources in a way that will enable the solution of the problem or successful achievement of the task	Critical literacy Operational literacy Reproduction literacy
Integration	To bring digital resources together in combinations relevant to the problem or task	Critical literacy Operational literacy
Analysis	To examine digital resources using concepts and models which will enable solution of the problem or successful achievement of the task	Critical literacy Information literacy
Synthesis	To recombine digital resources in new ways which will enable solution of the problem or successful achievement of the task	Critical literacy Reproduction literacy Operational literacy
Creation	To create new knowledge objects, units of information, media products or other digital outputs which will contribute to task achievement or problem solution	Critical literacy Information literacy, Reproduction literacy
Communication	To interact with relevant others whilst dealing with the problem or task	Critical literacy Social-emotional literacy
Dissemination	To present the solutions or outputs to relevant others	Critical literacy Social-emotional literacy
Reflection	To consider the success of the problem-solving or task-achievement process, and to reflect upon one's own development as a digitally literate person	

At the second level of the model is *Digital Usage.* At this level, the competencies developed in Level I are applied within specific professional or domain contexts (for example work, study, leisure), where the individual's existing digital literacy and the requirements of the problem or task shape the solution to the problem/task. In order to solve the problem or complete the task, Martin and Grudziecki (p. 258) stated that:

> the individual identifies a competence requirement. He/she may then acquire the needed digital competence through whatever learning process is available and preferred. He/she can then make an appropriate use of the acquired digital competence. The informed uses of digital competence within life-situations are termed here *digital usages*. These involve using digital tools to seek, find and process information, and then to develop a product or solution addressing the task or problem. This outcome will itself be the trigger for further action in the life context.

At level III, the *Digital Transformation* level, the digital usages attained in level II is expected to bring about change such as innovation and creativity, at the individual or group/organisational level. Martin and Grudziecki stated that transformation is not a necessary condition of digital literacy even though many digitally literate individuals may achieve transformative experiences. With levels of appropriate and informed usage, an individual would be considered digitally literate. Furthermore, the path to digital literacy development at each stage does not have to be sequential or entrenched within specific tasks. The pattern to developing digital literacy could be random where individuals draw on whatever skills is necessary including lower level knowledge and skills to develop or understand materials that are of higher order.

The implication of embracing the above framework or any other framework for developing digital literacy in students would be the need for it to be embedded in the curriculum. The two approaches adopted in schools to do this are usually:

1. teach exclusively in specialist ICT courses for the development of digital literacy, ensuring that it is not only technical skills that are taught but that tasks/activities are designed to provide opportunities to develop the conceptual components of digital literacy as well. The argument for this approach is that the technical aspect of using ICT could be taught more proficiently in these types of ICT-based courses. The argument against this approach is the non-authentic nature of the activities since it is not possible for ICT teachers to be knowledgeable across all discipline areas and to teach and assess across the different areas.

2. the integrated approach where every teacher implement the use of ICT in the teaching of his/her discipline area, for example, the science teacher would introduce *Inspiration* for concept mapping of key concepts for the topic on electricity and teach about the technical aspects of the software. Integrating ICT across the curriculum would require a whole school approach so that teachers are aware of the types of ICT that other teachers are using and that the teaching of skills to use particular software is coordinated. An argument against this approach is that many methods teachers do not feel adequately prepared to handle the plethora of ICT resources that are available for teaching and learning.

The second approach is the more popular approach, for example in Australia, the *Teaching Teachers for the Future*[6] initiative is a recent injection of nearly $8 million dollars into the ICT building capacity of pre-service teachers. Adopting this approach means that methods teachers need to be digitally literate in using the various technologies in his/her discipline area.

CONCLUSION

The development of digital literacy is an inevitable part of education in today's contemporary society. The chapter has proposed a framework for digital literacy and shown the interdependency of a multitude of literacies in it. While the notion of Prensky's (2001) 'digital natives' is being debated in academic circles, even those who oppose the concept would say that the young people today are better handlers of technology, particularly within the areas of mobile phones and social networking sites usage (Downes, 2005; Upper Yarra Community House, nd). Many of these young people would have a degree of digital literacy that they develop informally, as indicated by Ito et al. (2008, p. 1-2):

> In both friendship-driven and interest-driven online activity, youth create and navigate new forms of expression and rules for social behavior. In the process, young people acquire various forms of technical and media literacy by exploring new interests, tinkering, and "messing around" with new forms of media...Through trial and error, youth add new media skills

[6] http://www.altc.edu.au/ttf/

to their repertoire, such as how to create a video or customize games or their MySpace page.

While equipped with the technical capabilities indicated by Ito et al. (2008) and the ability to use mobile devices such as smartphones and tablets (iPads and other Android-based tablets) to connect with friends online and follow an online community with a particular interest (for example via *Tweeter* or *RSS* [7]feeds) or find their way around using GPS[8], many students would still need to be taught explicitly the cognitive and social aspects of using these devices and the applications that go with them. Schools should provide dedicated and real time in class to learn these aspects of digital literacy through the provision of authentic scenarios to critique and/or problems to solve. Students should be encouraged to track their own online activities over a period of time and keep a reflective journal or blog that assess their own activities and identify issues encountered or potential issues that could arise.

In addition, as most students are unfamiliar with educational technologies, they would need to be taught explicitly about these technologies, some of which are specific to discipline areas, and how to use them appropriately. Science teachers have expressed concerns about software programs that have complicated and non-user friendly interfaces as it often deterred students (and themselves) from staying focused on the tasks at the computer and spending too much time on the technical aspect of the software (Ng & Gunstone, 2003). Examples that these science teachers cited that were difficult to use technically included datalogging and *Flash* animation software. For both teachers and students, it is necessary to invest time to explore the software, including reading the manual and 'Help' tool when obstacles are encountered. Cognitive skills to evaluate the effectiveness of the software and to determine whether it is worthwhile using the software are necessary aspects of being digitally literate in using educational technologies. The implications of this for teachers are discussed in Chapter 7 of this book.

[7] Really Simple Syndication. It allows people to stay informed by retrieving the latest content from the sites that they are interested in.

[8] Global Positional System

Chapter 4

MULTILITERACIES: MULTIMODAL MEANS OF SUPPORTING SCIENCE LITERACY

INTRODUCTION

The influence of digital technologies on meaning making (the construction of understanding) in the last two decades has seen the concept of 'literacy' slowly evolving to the literacy plurality of 'multiliteracies' and the definition of language 'morphing into metalanguage' (Adwell et al., 2007). The term 'multiliteracies' was first proposed by the New London Group (1996) to highlight two arguments that are influenced by globalisation and technology: (i) the significance of cultural and linguistic diversity and (ii) the multiplicity of communication channels. The former argument asserts that as today's society becomes more linguistically and culturally diverse, meaning making differs according to cultural, social and professional contexts while the latter suggests that media and communication technologies enable meaning making to be increasingly multimodal and where the written-linguistic modes are integral of visual, audio, gestural and spatial patterns of meaning.

This chapter will focus on the concept of multiliteracies where the second argument of multiliteracies - the multimodal perspectives of representations (in supporting scientific literacy), will be emphasised. Before expanding on the term 'multiliteracies', it is worth distinguishing it from two other terms that seem similar in meaning to it. These are the terms 'multiple literacies' and 'new literacies'. It is not the intention of this chapter to tease out the philosophical and subtle differences between them. Rather, the intention is to review the broad meanings of these terms to position the use of the words multiliteracies, multimodality and multiple literacies in this book.

Multiple Literacies, Multiliteracies and New Literacies

Paul (2006) stated 'multiple literacies' as

> ...diverse sets of practices and semiotic systems to construct, acquire, communicate, and question knowledge as well as to create, analyze, and transform relationships among people and between people and institutions. (p.384)

This description of multiple literacies does not exclude the practices related to print literacy. But the affordances that technologies offer broaden the diversity of practices and semiotic systems. For Kellner (2004), who asserted the need to cultivate 'multiple literacies' into the structure of education in an era of technological revolution, the diversity of practices and semiotic systems, as stated by Paul (2006), are those of media literacy, computer literacy and multimedia literacy. Multiple literacies could therefore be viewed as the different literacies within a domain, for example, within the domain of *multiliteracies* are the multiple literacies of linguistic, visual, audio, gestural and spatial ways of meaning making (New London Group, 1996) while within the domain of *digital literacy* as described in Chapter 3, there are multiple literacies such as information literacy, cybersafety literacy, branching literacy, critical literacy, multiliteracies etc (see Figure 7 in Chapter 3). While multiple literacies is used interchangeably with multiliteracies sometimes (for example, Fortuna, Henderson, McLuckie, Rodrigues, Syme-Smith, Taylor & Williamson, 2010), this book concords with this interchangeability because there are multiple literacies associated with multiliteracies. Multiliteracies itself could also be used as a singular entity, being one of the multiple literacies of digital literacy. Within the context of this book, the relationships between multiple literacies, multiliteracies and digital literacy in ICT- enabled learning environments are shown in Figure 1. As will be discussed in a later section of this chapter, the multiple literacies of multiliteracies in science learning are related to the multiple modes of representations that help students develop scientific literacy.

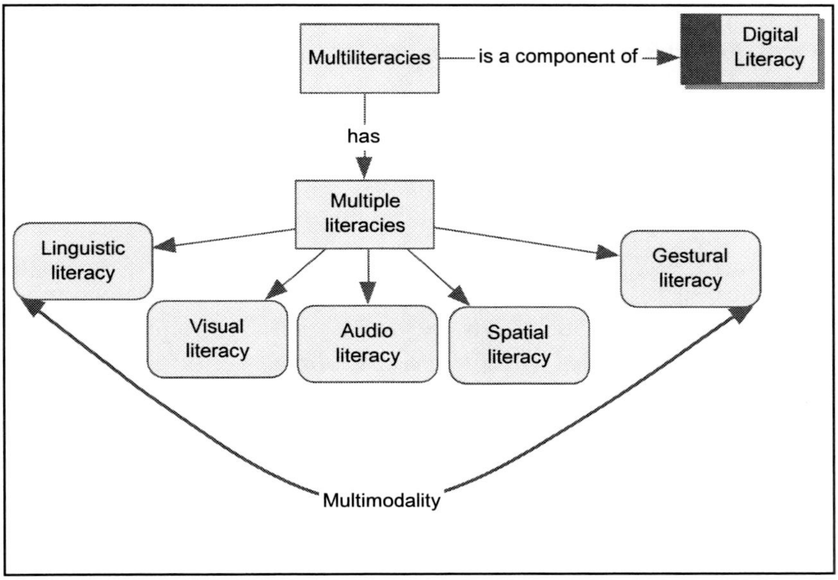

Figure 1. Relationship between multiple literacies, multiliteracies and digital literacy.

'New literacies' is a relatively new concept in the literature. According to Australia's newliteracies.com.au website, new literacies are digital literacies characterised by SMS (short message service), MMS (multimedia messaging service), social networking activities and mobile technologies such as mobile phones, smartphones and tablets. The website describes 'new literacies' as a combination of "letters, symbols, colours, sounds and graphics to extend language and the ways we communicate". Lankshear and Knobel (2003) attributed new literacies to the new types of knowledge associated with "digitally saturated social practices". They stated that a shift in mindset is necessary to embrace 'new literacies' education in order to take into account the changes the world has experienced during the information technology revolution. They described new literacies (p. 16-17, cited in Martin and Grudzieckim, 2006) as:

> ... 'posttypographic' forms of textual practice. These include using and constructing hyperlinks between documents and/or images, sounds, movies, semiotic languages (such as ...emoticons ('smileys') used in email, online chat space or in instant messaging), manipulating a mouse to move around *within* a text, reading file extension and identifying what software will 'read' each file, producing 'non-linear' texts, navigating three dimensional worlds online and so on.

In their paper *Toward a Theory of New Literacies Emerging From the Internet and Other Information and Communication Technologies*, Leu, Kinzer, Coiro and Cammack (2004), drawing on Bruce (1997), Leu (2002) and Reinking's (1998) work, stated that a more precise definition of the characteristics of new literacies would be difficult to achieve because of the continually evolving nature of technology and that as new technologies appear, new literacies may emerge as well. Similar to the argument of Lankshear and Knobel (2003), they attributed new literacies to new emerging technologies such as web logs (blogs), video editors, web browsers, web editors, e-mail, presentation software, instant messaging, plug-ins for web resources, listservs, bulletin boards, avatars, virtual worlds, and many others (p. 1571). They described new literacies as:

> The new literacies of the Internet and other ICTs include the skills, strategies, and dispositions necessary to successfully use and adapt to the rapidly changing information and communication technologies and contexts that continuously emerge in our world and influence all areas of our personal and professional lives. These new literacies allow us to use the Internet and other ICTs to identify important questions, locate information, critically evaluate the usefulness of that information, synthesize information to answer those questions, and then communicate the answers to others. (p.1572)

Both Lankshear and Knobel (2003) and Leu et al. (2004) advocated new literacies as adapting literacies that will continually evolve as new technologies emerge. In this respect, a digitally literate person should be able to adapt to new and emerging technologies quickly and pick up new semiotic language for communication as they arise, easily. Leu et al. (2004, p. 1589) have further gone on to identify 10 central principles of new literacies emerging from the Internet and other ICTs as:

1. The Internet and other ICTs are central technologies for literacy within global community in an information age.
2. The Internet and other ICTs require new literacies to fully access their potential.
3. New literacies are deictic.
4. The relationship between literacy and technology is transactional.
5. New literacies are multiple in nature.
6. Critical literacies are central to the new literacies.
7. New forms of strategic knowledge are central to the new literacies.

8. Speed counts in important ways within the new literacies.
9. Learning often is socially constructed within new literacies.
10. Teachers become more important, though their role changes, within new literacy classrooms.

Many of these principles are shared with those underpinning the digital literacy framework proposed in Chapter 3. The five notable similarities are (i) statement 2, the Internet and other ICTs requiring new literacies/digital literacy to fully access their potential (ii) statement 5, the multiple nature of new literacies/digital literacy (iii) statement 6, critical literacies are central to new literacies/digital literacy (iv) statement 9, learning is socially constructed within new literacies/digital literacy contexts in digitally-enhanced learning environments and (v) statement 10, teachers become more important in embracing new literacies/ digital literacy in the classroom.

Whether it is digital literacy, multiple literacy, multiliteracies or new literacies, a common feature in the descriptions of the terms is the general acceptance of the need to conceptualise a paradigm that sees the shift of text-based literacy to a multitude of literacies that are enabled by ICT.

Multiliteracies: Multimodality in Support of Scientific Literacy Development

Modes of Representation

Reading digital science content on a website or from a CD often involves engaging with information that are symbolic, graphical and pictorial. Science content is also accessible and downloadable in audio, video and multimedia formats. These varied representational formats are made possible by ICT's multimodal capacities. Learning from these representational possibilities would however, require interpretive skills beyond those emphasised in print-based literacy (Mulhearn & Hill, 2007). Paul (2006) drawing on Tyner (1998) stated that 'literacies of representations' include multiple literacies such as information literacy, visual literacy, media literacy, mathematical literacy, sign literacy and print literacy. These literacies are necessary to analyse, interpret and understand the meanings created. While this chapter will not be discussing these forms of literacies, it will look at the multimodal means of representing science concepts that are afforded by ICT and provide examples of how multimodal representations could help students develop scientific literacy.

Knain (2006, p. 659) spoke of multimodality in science as:

> *Multimodality* is a characteristic of scientific texts, and information and communication technology (ICT) of course enhances this aspect of text significantly. Multimodality should be part of the model both as an aspect of the competence of science literacy that students are expected to acquire, but also as a valuable tool for developing science literacy, by enabling transformations across different modes of representation and different contextual presuppositions, in particular between everyday and scientific contextual frames. *Transformations* should be part of the understanding of this dimension also as they connect multimodality to the origin of conceptual knowledge in action and culture.

The different modes of representation alluded by Knain are derived from the multiliteracies work of the New London Group (1996) who, as indicated at the beginning of the chapter, suggested that meaning making involve linguistic, visual/spatial, audio, gestural (or embodied in the case of science role-playing and conducting experiments) modes of representations. Hence multimodality is a combination of two or more modes of representations. These modes concord with those of Bruner's (1960) who proposed that children develop knowledge through the interactions of three distinct modes of representing the world: enactive (action-based), iconic (image-based) and symbolic (language-based) – see Figure 2. He suggested that children think through these modes because people around them interact and perform tasks through actions and the use of pictures and words. Unlike Piaget's stages of cognitive development (1955, 1972), Bruner's cognitive development is not linear and the modes of representation are only loosely sequential as they integrate into one another. In Bruner's mind, language-based tools are considered the most important in enabling abstract thinking and reasoning through symbolic modes of representations. In mapping the multimodality perspective of multiliteracies of the New London Group (1996) into Bruner's (1960) modes of representations (see Table 1), the audio literacy of multiliteracies has a less obvious fitting into Bruner's model, but would seem to fit into the verbal aspect of his symbolic mode of representation where 'text' could be written or verbal. Audio literacy would include other sounds, songs and music and how they are expressed. In both the New London Group and Bruner's modes of representations, multimedia resources would be multimodal in nature and combines two or more modes of representations.

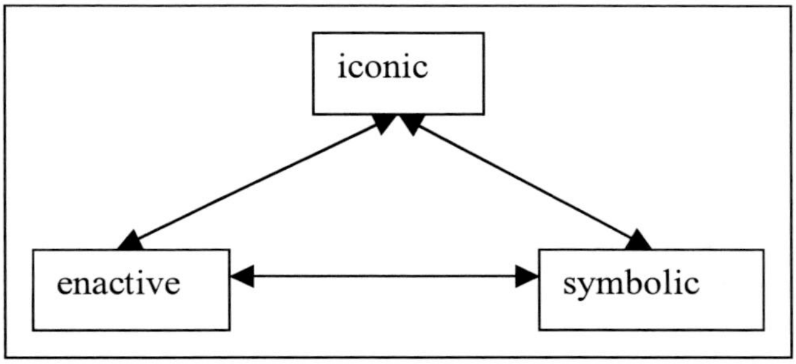

Figure 2. Bruner's modes of representations.

Table 1. Modes of representations of the New London Group (1996) and Bruner (1960)

New London Group (1996) Multiliteracies	Bruner (1960) Modes of Representations
Gestural (or embodied)	Enactive
Visual, Spatial	Iconic
Linguistic, audio	Symbolic
Multimodal (combination of two or more modes of representations that are gestural, visual, spatial, linguistic or audio)	Multimodal (combination of two or all of enactive, iconic, symbolic modes of representation)

While Bruner viewed each mode as being dominant at different times during a child's development, learning is essentially represented by a combination of these modes. In science learning, enactive representations can be manifested through role-plays or manipulating objects through experimentations and modelling; iconic representations can be demonstrated using pictures, videos, simulations, illustrations and diagrams while symbolic representations can be shown through text (written or verbal), numbers, graphs, tables and charts. Multimodal communication reinforces how thinking involves transforming these representations and understanding the connections between the different knowledge representations.

In science education, learning involves the exploration and interpretation, as well as production of multiple representations of the concepts under study. It also involves the ability to create connections between the different modes of representations in order to develop and demonstrate reasoning processes and understandings (Ainsworth, 1999, 2005; Bardini, Pierce and Stacey, 2004; Prain and Waldrip, 2006; Stylianou and Pitta-Pantazi, 2002; Waldrip, Prain &

Carolan, 2006). Helping students to understand concepts requires both the teacher and students to work with the students' representations that make what they understand explicit. Depending on the nature of the concept under study, some modes of representations lend themselves better to representing a concept, for example a text-based description of abstract concepts like light rays and how they bend in different medium is perhaps less easy to grasp than a visual or animated model that is multimodal. As different modes of representations have different strengths and weaknesses in demonstrating associated meanings, students need to be shown how to use the different modes of representation to fit the purpose of the science learning and to demonstrate effectively their understandings of the concepts learned (Prain & Waldrip, 2006; Waldrip, Prain & Carolan, 2010).

ICT-Enabled Modes of Representations in Science

In science learning, modes of representations afforded by digital technologies are shown in Table 2. These include symbolic (written and verbal), iconic (visual and spatial/three-dimensional), enacted (gestural/embodied) and a combination of modes (multimodal) of representations. The table shows the different modes of representations that could be presented or captured by digital technologies. In this book, multimodal refers to the simultaneous use of two or more modes of representations to convey science concepts or to represent a student's science understanding and reasoning. Multimodal representations of concepts that demonstrate students' understanding could be captured on desktops or handheld computers such as the personal digital assistance (PDA). Examples of PDA-captured multimodal representations of students' understanding are shown in Figure 3. Figure 3(a) shows a six-year old child making use of the voice recording function of the handheld computer to record a verbal explanation (symbolic) of the drawing (iconic) of a butterfly and the environment it is in. Figure 3(b) shows a cell drawn by an eleven year-old student, where iconic (image of cell) and symbolic (text) modes of representations are displayed.

In science, a practical report containing diagrams of equipment, images, videos or graphs of results and text to describe and discuss results is representing understanding in a multimodal manner. An example is shown in Figure 4. It contains components of a university primary pre-service teacher's investigation report of air pressure and how many plastic cups could a blown-up balloon hold around it. The figure shows a part of her report that contained photos, written text and a table summarizing her attempts. Had a video been made to show the process of blowing up the balloon and the 'sticking' of the cups to the balloon, gestural/enactive mode of representation would have been demonstrated.

Table 2. Modes of representations in science enabled by digital technology and software

Mode of representation	Demonstrated in	Examples of digital technology/software enabling learning or creating
Written	Text	Word, Pages
	Worksheets; puzzles	Word, Pages; Hot Potatoes; SurveyMonkey; Quia; Wildform; QuizCreator
	Assignments; projects	Word, Pages; Web search for information
Verbal	Oral presentations	Audio/video recording; podcasts
Visual	Drawings/diagrams	Drawing software e.g. DrawPlus, SmartDraw and ACD/ChemSketch; digital camera to capture diagrams; scanner;
	Concept maps	Inspiration; Kidspiration; iMindMap, FreeMind
	Tables and graphs	Excel; Numbers
	Animations/simulations	Flash, java applets; simulated experimental work; GoAnimate;
	Presentations	PowerPoint; Keynote; Prezi
Embodied	Role play; science drama	Digital. video recorder; digital camera
	Modelling (hands-on)	Digital camera and or video recorder
Spatial/3-D modes	Models (visual)	Visualization software e.g. VDN and Molekel
Multimodal	Experimental work & report	Word/Notes for text; Excel/Numbers for graphs; digital camera to capture results; video recording a process
	Concept development	Range of Biology, Chemistry and Physics software online and offline

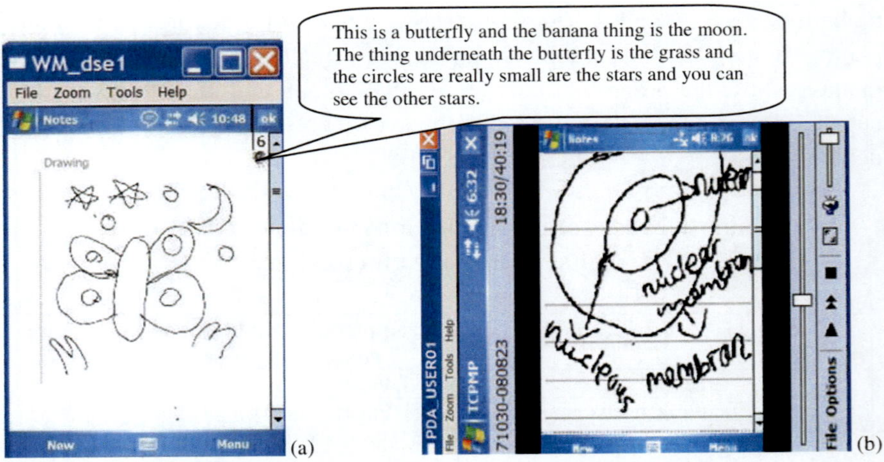

Figure 3. Examples of multimodal representations (a) image and verbal (voice recorded) and (b) image and text.

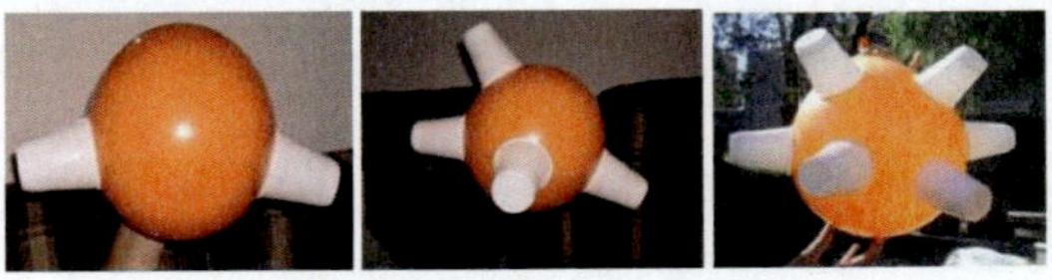

Air Pressure: Balloon and Cups.

Results:

Attempts	**Number of cups 1st puff**	**Number of cups 2nd puff**	**Number of cups 3rd puff**	**Number of cups 4th puff**	**Number of cups 5th puff**	**Total number of cups on inflated balloon**
1	0	2	0	0	0	2
2	2	0	2	0	0	4
3	2	2	2	-5	0	1
4	2	2	2	0	Balloon burst	0
5	2	2	2	0	0	6

Figure 4. Multimodality in a science report.

Multiple Representations

The term 'multiple representations' refers to the re-representing of the same concept under study in different ways through different modalities.

Below are two examples to illustrate the multiple representations of 'Changes of State' and 'Photosynthesis and Respiration'.

Example 1- Changes of States of Solids, Liquids and Gases

This example on changes of state of solids, liquid and gases is adapted from Ng and Nicholas (2011, p.82-83). In studying about changes of states that embrace the *Particle Theory*, the multiple representations used to show the strengths of the bonds between each of the states of matter and how they are affected by heat could include:

(i) role play (enactive/ gestural representation). Students themselves experience 'being' water particles in this example of a role play. In the solid (ice) state, by linking each other's arms at the elbows, students will automatically fall into a neat orderly arrangement because the 'bonding' (arm-linking) between the students is strong and movement is limited. The only room for movement is the vibration of each student (particle) while fixed in its position. Students could demonstrate this by movement on the spot. In the liquid (water) state, the elbow-linking changes to hand-holding between the students. Hand-holding 'bonds' are less strong with a greater degree of freedom for students to move in front of or behind each other (representing water molecules sliding over each other) while still holding hands. The transformation to the gaseous state is represented by the students releasing their hand-holding state to break completely free from each other and spreading themselves around the room, including some students possibly going out of the classroom door if it is opened. The freedom to move around and the disappearance of the students outside the classroom door in the gaseous state represents the dimension of understanding that the forces of attraction between gas particles are almost non-existent and that the particles spread themselves to take up all the available space. The changes of state require heat to provide energy for the water molecules to break away from each other. This could be represented by a red piece of cloth as 'flame' that fans each state to enable the change from one state to another to occur. The reverse transformation would take place if a white or blue piece of cloth signifying coldness is used. A video recording of the role play with voice over explaining the processes could also be made. Multimodal short videos such as this could be made using *MovieMaker* or *iMovie* software.

The role play could be extended to simulate sugar dissolving in water and dilute and concentrated solutions.

(ii) *creative writing (symbolic/ text representation).* The student could write a creative story on the adventures of a block of ice and the transformation it undergoes in different environments. The student will need to ensure that the concept of energy is demonstrated and how this affects the strengths of the bonds holding the water molecules in the different states.

(iii) *diagrammatic representation (multimodal representation).* Students could construct a labelled diagrammatic summary of the changes of states, their relationships with energy and the arrangement of particles in the different states. An example is shown in Figure 5. The modes of representations in this figure are text (symbolic) and images (iconic).

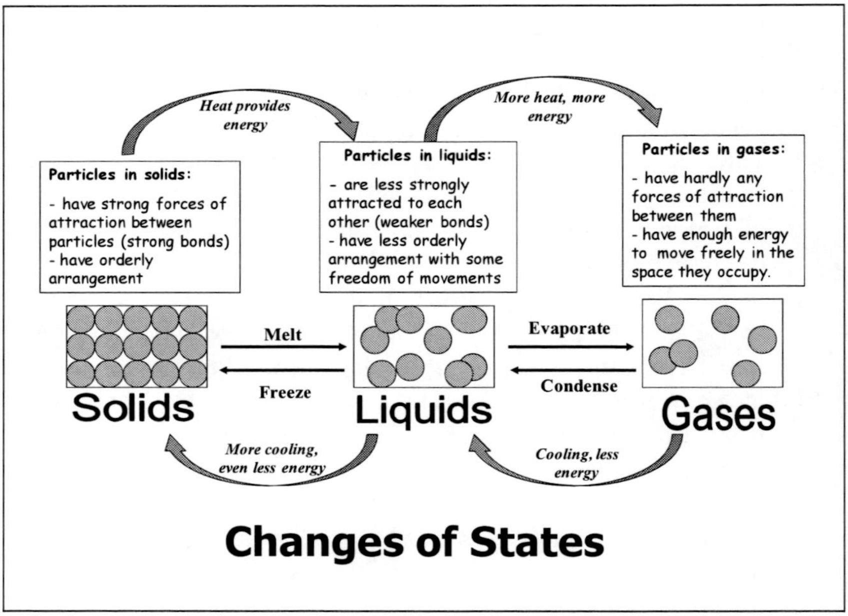

Figure 5. Multimodal representation of the *Changes of States.*

(iv) *experimenting (multimodal representation) and explaining (symbolic representation either through written or verbal text).* As an example, the students could conduct the activity of 'making clouds in a bottle'. This would require students to carry out the activity, observe, video record and explain the phenomenon that took place inside the plastic bottle containing a little water and a burnt match. The modes of representations are gestural (enacted) in carrying out the investigation and verbal or written text (symbolic) in explaining the phenomenon in the bottle.

Example 2 – Photosynthesis and Respiration

Photosynthesis and respiration are two concepts usually studied together at middle to upper secondary year levels. Advanced concepts such as ATP[1] as energy units and the biochemical reactions that occur within chloroplasts as light dependent (light reactions) or non-light dependent (dark reactions) processes are not covered at these beginning stages of learning about photosynthesis. Multimodal and multiple representations of understanding of these concepts are shown below.

(i) *Diagrams (visual/iconic).* Diagrams of plant and animal cells and their internal components such as chloroplasts, stoma, chlorophyll and mitochondria are iconic representations that will help students visualize structures required for the processes of photosynthesis and respiration to take place at the microscopic level.

(ii) *Illustrations of chemical reactions (symbolic/text and symbols).* This could be in the form of words or equations. For example, the word equation for photosynthesis is

$$\text{carbon dioxide} + \text{water} \xrightarrow[\text{chlorophyll}]{} \text{glucose} + \text{oxygen}$$

and the formulae-based chemical equation is

$$6CO_2 + 12H_2O \xrightarrow[\text{chlorophyll}]{\text{sunlight}} C_6H_{12}O_6 + 6O_2$$

At a higher level of learning, introducing students to the molecular structures of the elements (oxygen) and compounds (water, carbon

[1] ATP is adenosine triphosphate

dioxide and glucose) involved in the processes of photosynthesis and respiration would cater for the more able students. The structures (visual/iconic representations) for the different components involved in the processes are shown in Figure 6.

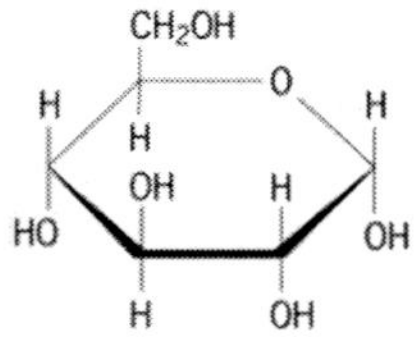

H-O-H O=C=O O=O

glucose **water** **carbon dioxide** **oxygen**

Figure 6. Structures of glucose, water, carbon dioxide and oxygen.

(iii) *Videos and animations (multimodal).* There are many websites that explain the processes of photosynthesis and respiration. A website that is suitable for secondary school students is the neok12.com website where all videos have been reviewed by K-12 teachers. Videos and animations contain images, text and voice over explanations, hence they are combinations of iconic/visual and symbolic/verbal representations.

(iv) *Concept mapping (iconic/visual and symbolic/text).* Concept maps are graphical tools for organizing and representing knowledge. They include representing key concepts in nodes and showing the relationships between them by a connecting link and linking word(s) between two concepts (Novak & Cañas, 2006). Key concepts could be given by the teacher or created by the students themselves. Figure 7 shows two concept maps using a set of keywords provided by the teacher for the process of respiration. The concept maps were constructed using the concept-mapping software *Inspiration*. Figures 7(a) and (b) show that there is not just one way of representing the same idea, topic or set of keywords in concept maps. Each of these representations provides insights into how well students are connecting key concepts studied.

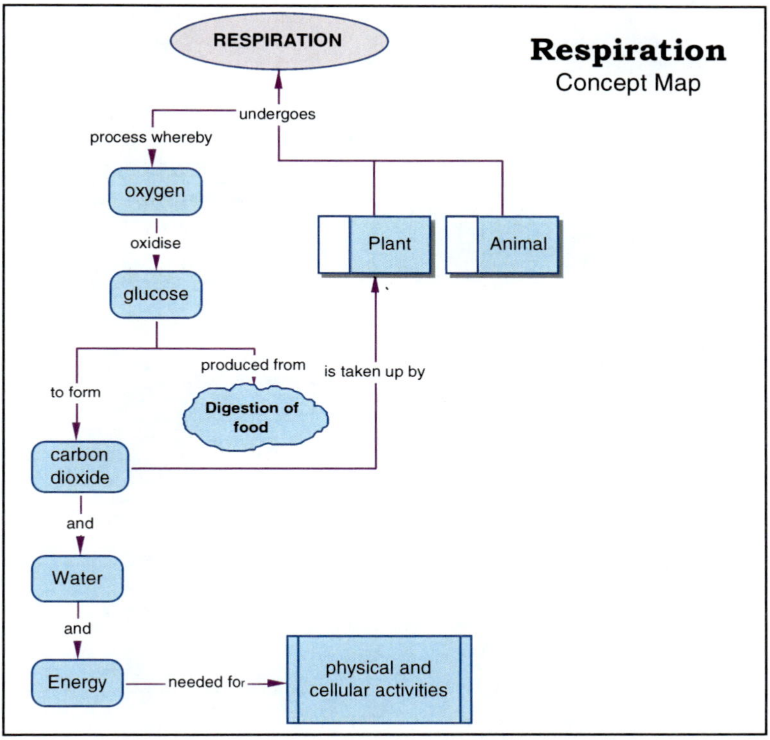

(a)

Figure 7. (Continued)

(iv) The linking word 'oxidise' used in the first concept map (Figure 7a) indicates a slightly higher level of understanding of the process. Similar concept maps could be constructed for the process of photosynthesis. In taking the learning to a higher level, students construct a composite concept map of photosynthesis and respiration showing the interrelationship (similarities and differences) between the two concepts using keywords associated with each process. An example is shown in Figure 8. Concept maps are a quick and useful way of providing a visual overview of what students are thinking. Different ways of using concept maps are discussed in Chapter 6.

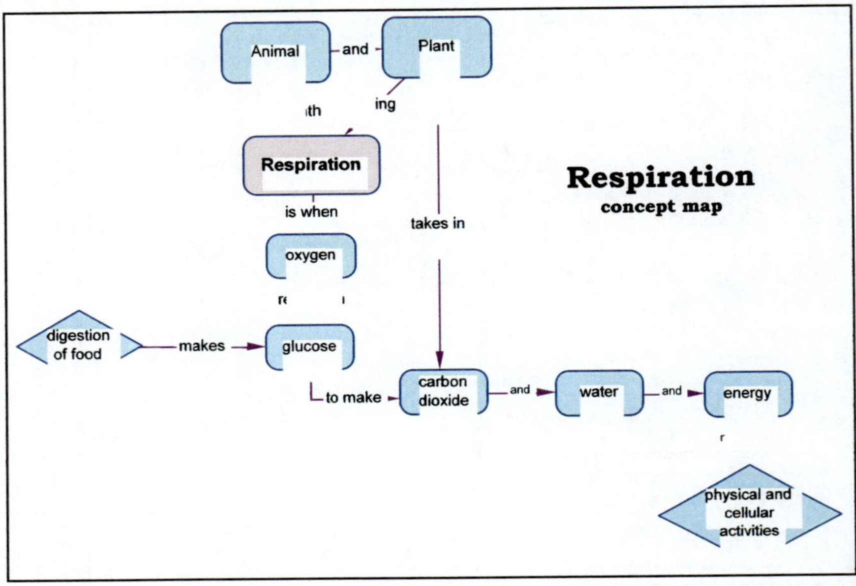

(b)

Figure 7 (a) and (b). Two versions of concept maps for the concept of *Respiration.*

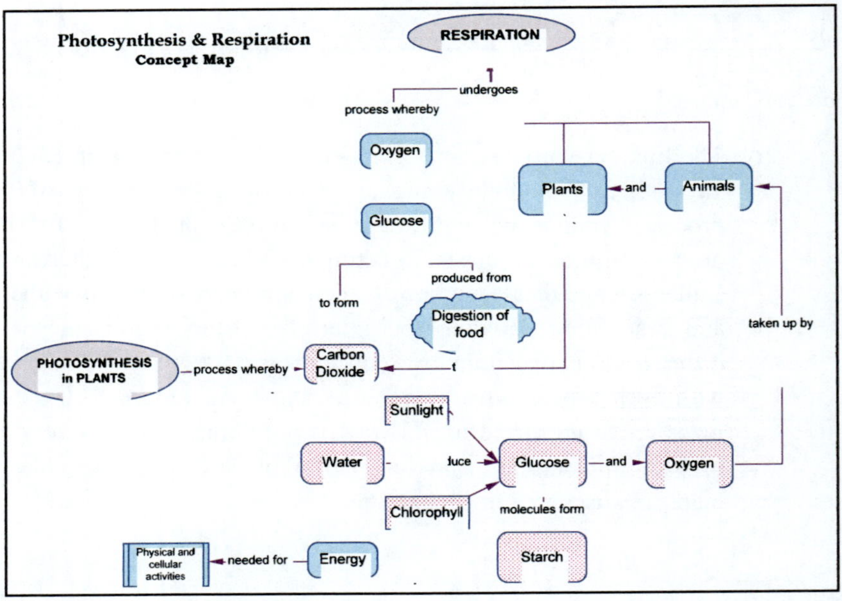

Figure 8. A composite concept map for *Respiration* and *Photosynthesis.*

CONCLUSION

Russell and McGuigan (2001) stated that in order to learn, learners need to be provided with opportunities to construct a variety of representations of a concept through which they refine their representations in different modes as they make more explicit their conceptual understandings. Nuthall (1999) proposed that in order to establish more permanent understandings, learners will require multiple exposures to the same concept. In accordance to these statements, this chapter has discussed the significance of the use of modes of representations in supporting scientific literacy development. It has distinguished between multimodal and multiple representations and demonstrated examples in the use of single and combined modes (multimodal) of representations. It has provided examples to show what it means to use multiple representations - the re-representing of the same concept in different ways to achieve more permanent understandings (Nuthall, 1999). Providing opportunities for students to learn from or to demonstrate understandings in a variety of modes of representations and in multiple representations through re-representing meanings are essential activities for developing scientific literacy. These opportunities assist students to refine their science understandings and allow teachers to probe deeper into their students' learning in terms of the accuracy of the representations of the concept under study. Hence teaching students to represent science knowledge and understanding in different modes to foster the development of scientific literacy is an important aspect of science education. As multimodal representations are enabled by ICT, being digitally literate would help the student learn better from a variety of ICT-based resources and in selecting and using appropriate applications to represent their understandings in multiple modes and through multiple representations.

The chapter has highlighted the multimodal perspective of multiliteracies (New London Group, 1996) and mapped its various literacies of visual, spatial, audio, linguistic, gestural and multimodal into Bruner's (1960) enactive, iconic and symbolic modes of representation, demonstrating general agreement in the modes of representations for learning between the two groups of researchers. As the terms of visual, audio, spatial, linguistic and gestural modes of representations are more familiar to the layman, I will use them in describing modes of representations for the rest of this book.

Chapter 5

A Framework for Empowering Scientific Literacy Through Digital Literacy and Multiliteracies

Introduction

In the preceding chapters, I have discussed the three main components of this book: scientific literacy, digital literacy and multiliteracies. This chapter aims to tie these components together to develop a framework for the empowering of scientific literacy development through digital literacy and multiliteracies, the latter being the multimodality of representations enabled by the affordances of digital technologies.

Digital Literacy Empowering Science Learning

There are similarities between digital literacy and scientific literacy, that is, a set of cognitive skills that are common to the development of both literacies. These are skills related to critical thinking and research. For digital literacy, this include being able to search, locate, assess (analyse and evaluate), decide (select for the most appropriate information), plan and synthesise new ideas and meanings from the information researched. In searching for information, an understanding of the cultural, ethical and social issues associated with the use of ICT in conducting the research, is part of being digitally literate. Critical thinking skills such as relating theory to practice, seeking evidence-based conclusions, asking appropriate questions and linking ideas together are necessary to research for new information and create understandings from it. Critical literacy skills are also required in the assessing of information and resources to ensure that materials are relevant, accurate and unbiased in both the content and the way they are presented.

From the 'scientific method' perspective of scientific literacy, many of the critical thinking and research-based skills needed to build this knowledge are similar to those of digital literacy (see Table 1). For example, in conducting experiments, critical thinking and research skills are related to the designing and conducting of an experiment, seek and locate (that is, observe and gather experimental) data, assess (analyse and evaluate) the data gathered, draw conclusions, synthesise meanings from the data and choose the most appropriate medium to display and communicate the findings. At the non-experimental level of scientific literacy, researching on the Internet for content information or information that addresses an issue of science in society means that individuals equipped with digital literacy skills and knowledge would be empowered to undertake the task(s) effectively.

In both digital and scientific literacies, an understanding of the ethical, social and cultural issues pertaining to the respective literacy development and the ability to adopt both critical thinking and critical literacy in evaluating information and in creating new ideas are essential. Plagiarism, the copying of other people’s work would be an ethical issue common to both literacies. Communicating is a social practice of both literacies. In scientific communities, scientific content and explanations, theoretical proposals and/or result findings from experimental work are communicated via email, web-based submission systems, journals (many articles are published online first), blogs, wikis, videos and on web pages that are either personally created or are part of an organisation’s website. In digital communities, communication of a variety of topics of interest takes place via email such as listserve, blogs, wikis, videos, slide sharing and social network sites. The cultural dimensions associated with digital literacy involve the understanding of the ‘rules’ when accessing online sites and the use of languages that are distinctive to the different platforms. For scientific literacy, components that are cultural to a scientific practice include evidence-based principles, the scientific method and scientific language.

The similarities in skills/knowledge for both literacies would mean that digitally literate individuals should be able to make use of digital literacy skills/knowledge to support their development of scientific literacy and vice versa. It means that time spent on developing digital literacy capabilities should be seen as a means of empowering the development of scientific literacy rather than as extra work.

Table 1. Similarities in the set of critical thinking and research-based skills for digital and scientific literacies

Digital literacy	Scientific literacy
Seek and locate for data	Seek and locate information data Observe and gather experimental data
Analyse and evaluate data gathered	Analyse and evaluate data gathered
Synthesise meanings from the data	Draw conclusions from the data
Communicate using appropriate medium	Display and communicate results using appropriate medium
Understand ethical, social and cultural issues associated with ICT use for learning and socialising	Understand ethical, social and cultural issues associated with science experimentations and communication
Adopt critical literacy skills to assess information and resources as well as in synthesising new content from researched materials	Adopt critical literacy skills to assess findings as well as in drawing conclusions and communicating the findings

There are other reasons for why digital literacy could empower science learning. One of them is that it assists students to learn more effectively from the range of ICT- enabled affordances that are available to motivate learning and to enable better understanding of science concepts. Another reason is that it lessens the working memory's cognitive load while learning science in an ICT-enhanced learning environment. These points are elaborated below.

ICT-Enabled Affordances in Science Education

Reimann and Goodyear (2004) proposed five general ways in which ICT can support successful learning. These are (i) increasing motivation (ii) providing highly interactive experience and rich feedback to engage with learning (iii) providing tools that demonstrate what has been learned (iv) providing means for communication and collaboration and (v) catering for differences in learning. In science education, Webb (2005) stated that the affordances offered by ICT benefit science learning by (i) promoting cognitive development (ii) enabling science to be related to students' real life experiences (iii) increasing students' self-management of their own learning and (iv) facilitating data collection and presentation . The points made in this paragraph by Reimann and Goodyear (2004) and Web (2005) are integrated and expanded below to demonstrate how

the use of ICT and being digitally literate could empower the learning of science.

Increasing Motivation

The motivational impact of technology, including mobile technology, on students' general learning as well as in science learning has been well documented in the literature of the last two decades (for example, Crook, Harrison, Farrington-Flint, Tomás and Underwood, 2010; Dywer, 1994; Higgins, 2003; Keogh, 2011; Mistler-Jackson & Songer, 2000; Ng, 2008; Ng & Gunstone, 2002, Pedretti, Mayer-Smith & Woodrow, 1998; Pittard, Bannister & Dunn, 2003; Wallace 2002). Students are motivated to learn with technology as they have ownership and control over their own learning in terms of pace and the choice of content.

Providing Highly Interactive Experiences and Rich Feedback to Promote Scientific Literacy Development

In science education, the affordances that ICT offer through the various types of software mean that information and learning materials can be delivered in a variety of modes for learning (see Table 2, Chapter 4). Multimodal representations in science learning are important due to the abstract nature of many of the concepts that students have to learn, particularly in chemistry and physics. Multimodality, as defined in Chapter 4, is the simultaneous representations of two or more modes in one representation to convey the intended meanings. For example, at the *Edheads*[1] game website for years 5-7 students studying about simple and compound machines, the use of images, text and animations to assist the students to visualise the working of a simple machine as an entity of a compound machine is shown. With this learning resource, a digitally literate student should be able to link the different

[1] http://www.edheads.org/activities/simple-machines/index.htm

modes of representations across multiple representations of the same concept when learning the science (Prain & Waldrip, 2006; Waldrip, Prain & Carolan, 2006) behind simple machines. The interactivity of the *Edheads* simple machine website means that students are able to go into authentic situations such as the bathroom, bedroom, kitchen or garage to identify the simple machines present in each of these areas. Having selected an area to explore, the student will need to know that to select for a simple machine, (s)he will have to roll the mouse over the items (for example bathtub, faucet etc) until one of them lights up. (S)he then proceeds to answer the questions displayed. Understanding the rules of the game is part of being digitally literate. For example, the student is allowed two tries to answer a question, where feedback is provided at each try. When the questions are answered, an explanation of the type of simple machine that the student has selected is given. For example in Figure 1, the visual (which is animated) representation demonstrates that a toilet paper roll on a roller is a wheel and axle type of simple machine. This multimodal representation is comprised of visual (picture and animation) representations that are reinforced by the text-based representation to help the student better understand the concept of a wheel and axle simple machine. The interactivity of the website with feedback provided at each level of interaction allows the student to work at his/her own pace to explore a series of authentic situations that they are able to relate to (Ng, 2008). Hence, ICT-enabled game-like quiz such as *Edheads* promotes scientific literacy development cognitively and motivates students' self-management of their own learning (Ng, 2008; Webb, 2005).

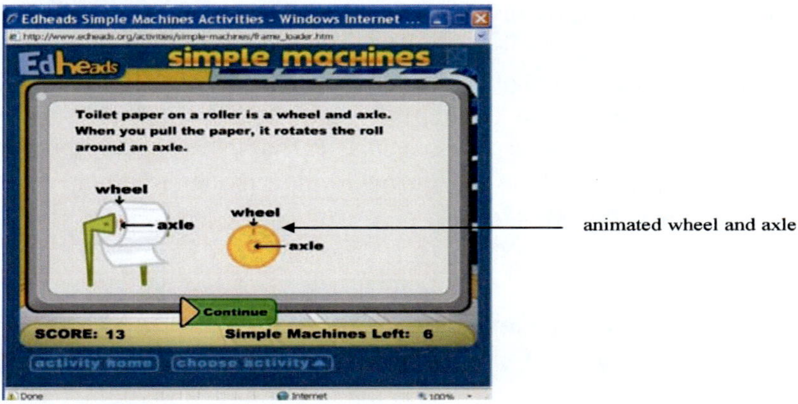

Figure 1. Snapshot of a multimodal representation from the *Edheads* simple machine page at http://www.edheads.org/activities/simple-machines/frame_loader.htm.

Apart from cost-free interactive resources online, there are also software packages that provide interactive, self-directed learning opportunities. Drill and practice software packages provide instant feedback, some are adaptive so that the applications can readily adapt to the user's changing needs, desires, and environment. Content-based software such as the *Crocodile* interactive series will help students learn about theories of electricity or chemistry and allow them to participate in virtual laboratory work. Students would need to know how to navigate around in these platforms and learn the operational features in order to maximise on the learning, for example where to get chemicals or equipment. They would need to know about the simulated characteristics of the software in carrying out virtual laboratory work, for example a titration or designing and connecting a complex circuit. They would also need to understand the complexity and limitations of virtual laboratories. For example, in a virtual titration, the need to control the titration and view volume numbers change simultaneously. Both the technical and cognitive dimensions of digital literacy are involved in interacting with these types of software and the more familiar the students are with the software, the better the learning.

Measuring variables and tracking data over time are common experimental activities in science learning. Data logging equipment that include a range of probes are available for students to carry out these activities. Using these equipment at the operational level and being able to read and interpret the data shown on the screen would require quite sophisticated digital literacy skills in operating the equipment technically and in interpreting the data displayed cognitively. Investing time to build up students' skills to use these equipment efficiently is worth the time in order to ensure that the learning of the science will not be impeded by their usage. Data logging is not restricted to operating from desktops or laptops. There are interfaces and probes available for mobile devices, such as PDAs, to carry out experiments. Such setups are particularly useful for fieldwork activities where the learning is situated within real-life contexts.

Learning with Mobile Devices: Interactive Experiences in Situated Learning

Over the last decade, mobile devices have become sufficiently advanced technologically and affordable financially to move beyond their experimental status. The use of mobile devices in education has moved from pilot projects to institution-wide adoption (see for example, Finn & Vandenham, 2004; Ng & Nicholas, 2007; Ng & Anastopoulou, 2011; Palm, 2002; Perry, 2003; van 't Hooft, 2006). Mobile devices are handheld devices that are small-sized, portable tools with a miniature keyboard or touch input (with a stylus or the fingers). The term typically covers personal communication devices such as:

- mobile/cell phones and smartphones
- mobile computers such as PDAs, Palmtops, Netbooks, Smartbooks and tablets (for example tablet PCs, iPads, Android-based tablets)
- media recorders such as voice recorders, digital still and video cameras (also known as camcorders)
- media players such as the iPod series (Classic, Touch, Nano and Shuffle) and
- e-book readers.

While the portability advantage is often offset by restricted input methods, limited displays (due to screen size) and low rates of connectivity, it cannot be denied that mobile technologies have become a normal part of everyday living, working and learning. According to the 2009 Australian Mobile Phone Lifestyle Index, accessing the web for videos, music and information on mobile phones is now a regular and normal part of life in the 21st century. The pervasiveness of mobile technologies in everyday activities is supporting learning, recreation and employment in a seamless and unobtrusive manner. Klopfer, Squire, and Jenkins (2002) stated that features of mobile devices that provide mobility and ubiquity are its portability (small size and weight), social interactivity (enabling collaboration and data exchange), context sensitivity (capacity to respond to current location, time and environment), connectivity (ability to create a shared network) and individuality (personalization, ownership and responsibility for own work). Under these circumstances, collaborative learning is promoted through social interactions and learning is situated within authentic contexts.

Situated learning (Brown, Collins & Duguid, 1989; Lave & Wenger, 1990) is learning as it normally occurs, that is, it is embedded within activity, context and culture. Rogers (1997) called it "natural" learning although it is also known as incidental learning, unplanned learning and unstructured learning. Unlike teacher-directed classroom learning, situated learning is usually unintentional rather than deliberate. Situated learning with mobile devices could take place within formal instructions

(i) in the classroom, for example taking images of experimental data, video recording a process happening such as the electrolysis in a petri dish[2] , recording with a stopwatch application the time taken for a chemical reaction to stop or to check the individual's at-rest and after-exercise pulse rates while learning about the circulatory system and
(ii) during school-led excursions, for example to a science discovery centre, museum, power station or soft-drink bottling factory. At these centres, students could voice record descriptions of exhibits that they see or the processes of power production and/or the bottling of soft drinks. They could capture interviews with personnel in these places and learn about science-related careers. They could enter data directly into *Mobile Excel* and do simple statistical calculations, for example of the bottling process on the spot.

Mobile devices are ideal for bridging formal and non-formal learning. This is achieved through note taking, image taking (where permitted), audio recording of sounds (for example at the zoo) and short interviews with people. For mobile devices with access to wifi or 3G/4G networks, students could check for the location of the place of visit through GPS, or search the Internet for more information about a display object or a science terminology. Data collected on-site are taken back to class or home for further processing.

Situated science learning also takes place informally, for example after school hours in the bus or train, during weekends and holidays and at home. This could take the form of listening to a science lecture podcast, reading a science eBook or an e-article, finding out the meanings of terminologies encountered while reading the newspaper/magazine or watching the television, connecting with peers to work on a collaborative science project or just jotting down thoughts and ideas whenever they come along. These activities are facilitated by mobile devices with the appropriate applications installed (for example e-dictionary). Hence, being digitally literate with using mobile devices and their applications would empower students to learn science ubiquitously.

[2] Electrolysis in a petri dish: http://www.carolina.com/category/teacher+resources.do

Content Creation: Tools to Demonstrate What Has Been Learned

There are numerous online and offline tools that students are able to use to demonstrate their understanding of concepts learned. The most common way that students demonstrate what they have learned is through essay writing or answering questions using word processing packages such as *Word* or *Pages*. Within these packages are a multitude of features that many students do not take advantage of or use to demonstrate their understandings. For example:

- font styles and colours to highlight or shade sections of the document to emphasise important points
- inserting images, tables and charts to make the document become more lively and visual, particularly in writing up reports for laboratory experiments conducted
- using the *SmartArt* features such as Venn, pyramidal and cyclical diagrams or flow charts to show relationships or get concepts across in a diagrammatic form
- annotate using the 'new comment' function
- using hyperlinks to make the flow of the document smoother and to provide more information to the reader on science terminologies or concepts written in the document.

Other tools that students could use to demonstrate their understanding include presentation software such as *PowerPoint or Keynote*, *Flash*, *Prezi* and *Movie Maker or iMovie*. Digitally literate students who have knowledge and skills with these software packages would be able to distinguish between them and select the one that would best convey the learning that they have gained. Each of these software packages has their own strengths and limitations in presenting meanings. For example, it may be easier to insert a non-*YouTube* video (such as a video recording of a water-alarm constructed out of recycled materials) in *PowerPoint* than in *Prezi*, the latter has been set up for embedding *YouTube* videos where the individual could just copy and paste the source code from the *YouTube* site into *Prezi*. All of these software packages have multimedia capabilities and the choice would be dependent on the purpose and content of the presentation.

Visuals that are at-a-glance view of how well key concepts are linked and understood are concept maps. Concept mapping tools that are commercial are

Inspiration and *Kidspiration.* Examples of open sourced Web 2.0 concept mapping tools are *Mind42* and the Institute for Human and Machine Cognition's (IHMC[3]) *Cmap Tools*. These are free online mindmap collaboration tools where individuals can share and develop ideas with their peers online. IHMC (http://cmap.ihmc.us/download/) wrote about *Cmap Tools* as:

> IHMC CmapTools program empowers users to construct, navigate, share and criticize knowledge models represented as concept maps. It allows users to, among many other features, construct their Cmaps in their personal computer, share them on servers (CmapServers) anywhere on the Internet, link their Cmaps to other Cmaps on servers, automatically create web pages of their concept maps on servers, edit their maps synchronously (at the same time) with other users on the Internet, and search the web for information relevant to a concept map.
>
> CmapTools is used worldwide in all domains of knowledge and by users of all ages to graphically express their understanding. In particular, CmapTools is used in schools, universities, government organizations, corporations, small companies, and other organizations, both individually and in groups, for education, training, knowledge management, brainstorming, organizing information, among other applications. The collaboration and publishing features provide a powerful means for representing and sharing knowledge.

Other Web 2.0 tools that enable students to demonstrate what they have learned are wikis (for example at *Wikispaces*, *Wikipia*, and *GoogleSites*), blogs (for example at *Blogspot*) and glogs - online posters such as at *Gloster* where students can make their own interactive posters, mixing in images, text, music and video and sharing them with peers. Similarly, students could create a website in free hosting sites like yolasite.com and weebly.com to display the work that they have done for their science project.

The wide ranging set of tools available to students for learning means that they need to have a sufficient level of digital literacy to distinguish between these tools, select and use the most appropriate ones to achieve the outcomes that they wish to demonstrate. As discussed in Chapter 3, while young people are generally quite comfortable with ICT usage, they are usually unaware of educational technologies, hence these tools need to be taught so that the students can develop the capabilities to use them.

---

[3] Institute for Human & Machine Cognition at http://cmap.ihmc.us

Tools for Communication, Collaboration and Sharing

A significant part of Web 2.0 is its social web with online tools that allow people to

- network socially through *Facebook*, *Google+, Ning*, *Skype* and other similar social networks or video conferencing facilities. Students discuss science projects and homework with their peers and sometimes with their teachers using these facilities.
- read, write and edit to build science knowledge collaboratively such as through wikis, *GoogleDocs* and *cMaps*.
- share science-related questions, ideas, opinions and knowledge through blogs, slideshare, podcasts (via *iTunes* or *uTunes*) and videos on *YouTube*.

The increased capacity to network with more learners and experts via Web 2.0, together with the 'always on' capability of mobile devices means that learners are able to access more resources to enhance their science learning more frequently. In formal learning situations, the use of online learning management systems (LMS) would facilitate interaction, collaboration and sharing of content files and opinions. LMS that are open sourced include *Moodle*, *Edmodo* and *Yammer* as platforms that teachers and lecturers could build a whole course around. The privacy of students is protected as login usernames and passwords are used for these systems.

What has been demonstrated so far is the capacity that ICT offers in helping students to develop their scientific literacy. The affordances include the capacity to promote higher order thinking skills, cater for different learning styles through multimodality and enable self-paced learning. Working with a multitude of tools means the need to develop proficient digital literacy and the associated critical thinking to work with them. As ICT skills and knowledge are transferable, the more digitally literate the individual is, the easier it is to adopt new technologies to learn about the science content, to synthesise content to demonstrate understanding or provide a solution to a problem.

Another reason for why being digitally literate is important for developing science literacy in ICT-enriched learning environments is the need to focus the working memory of the mind on the science learning and not to overload it with technological inabilities that will distract the mind from the learning.

DIGITAL LITERACY LESSENS THE WORKING MEMORY'S COGNITIVE LOAD WHILE LEARNING SCIENCE WITH ICT RESOURCES

The human cognitive model of information processing, according to Moreno and Mayer (2000) has three components: sensory memory, working memory and long term memory. External stimulus and information detected by the eyes (visual) and ears (audio) flow through the sensory channels to the working memory where the information is processed, coded and stored in schematic form in the long term memory (see Figure 2). Work by Baddeley (1986), Paivio (1986) and Mayer (1997) showed that there are two types of sensory memories that detect visual and audio information and that the information is processed in separate sensory channels in the working memory. As shown in Figure 3, each channel process different types of information modalities that are either image-based (photos, animations) or audio-based (sound, narration). The working memory acts as a temporary storage for information received from the sensory memories. It processes and manipulates the information as complex cognitive tasks, for example reasoning and understanding - the processes of learning. It draws on the long-term memory's storage of schemas (prior knowledge) to assist with its processing of the information received. Processed information is then stored in the long-term memory as schemas. Hence a schema is anything that has been learned and acquired, and has been constructed by a number of interacting informational elements in the working memory.

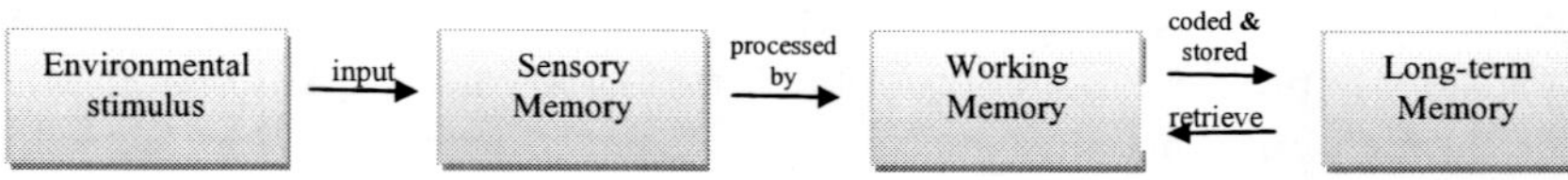

Figure 2. Human cognitive model of information processing (adapted from Moreno & Mayer, 2000).

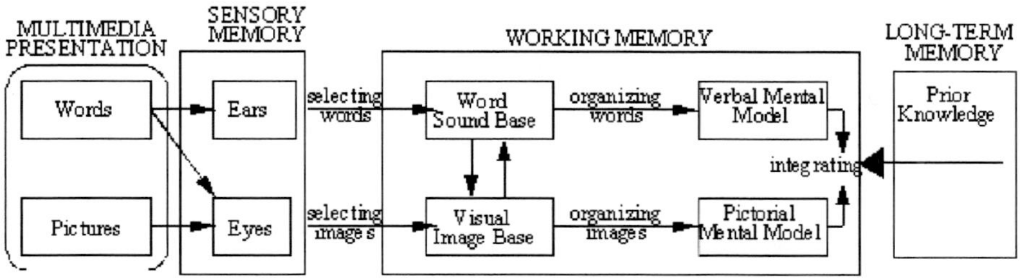

Figure 3. Cognitive theory of multimedia learning (Moreno & Mayer, 2000).

COGNITIVE LOAD THEORY

Research has shown that the working memory (also known as short-term memory) of the brain can process only a limited number of elements while learning (Miller, 1956). According to Miller (1956), the capacity limit is seven elements, called chunks, which could be retained for about 20 seconds before the information is lost (Sweller, 2009). These elements could be numbers, letters, words, sentences or other units. But many tasks would require more than seven chunks to be processed at one time, for example if the working memory has reached full capacity after reading seven sentences of scientific text, the individual will never understand the concept fully. What has been read is stored in long-term memory so that it can be drawn upon through the retrieval structures to make the necessary connections (Ericsson & Kintsch, 1995) as the reading proceeds. It is the knowledge accumulated in long-term memory (also known as prior knowledge) that determines an individual's intellectual degree, skills and abilities (Sweller, Merrienboer & Paas, 1998).

For the learner, when the technical skills and knowledge related to the technology at hand become very familiar and automatised, it will reduce the load on the working memory, enabling the student to focus his/her working memory on the task at hand rather than on the technology. The cognitive load theory (Sweller, 1988, 2005) states that there are three types of cognitive load:

(i) *intrinsic cognitive load.* This is the inherent level of difficulty associated with the complexity of the interacting elements of the instructional material that has to be processed simultaneously in the working memory. People have limited cognitive processing abilities and the number of elements (s)he can process depends on the level of expertise and the level of difficulty of the task. Instructors do not have control of this load but they can ensure that the learning materials contain appropriate number of elements so that the working memory can process the information to ensure that some learning could be achieved

(ii) *extraneous cognitive load.* This is dependent on how the instructor presents the learning material to the student and is the load imposed by poor design of the instructional materials, hindering students' understanding. Well designed learning materials would reduce the extraneous cognitive load and increase the capacity of the working memory

(iii) *germane cognitive load.* Germane cognitive load has also been implied as "effective' cognitive load. It is the load imposed by instructional materials that fosters the process of learning, for example motivational learning materials. Germane load is relevant for learning whereas extraneous load is not. Good pedagogy means that teachers design instructional materials to reduce the extraneous load and increase the germane load.

For a student learning science with the use of technology, by being skilled technically with working the features of the technology, (s)he does not have to split his/her attention between the technology and the concepts being studied. This would reduce the extraneous load, increasing the working memory's capacity to focus on processing the material to be learned. Depending on the science task and the ICT being used for it, the attention that the student needs to give to the technical and cognitive aspects of digital literacy would vary. For example, undertaking quizzes such as multiple choice questions would require little technical skills and the working memory could focus almost entirely on the content of the quiz to select the appropriate responses. On the other hand, the cognitive load associated with using data logging equipment to collect experimental data could be substantial. It would require the working memory to deal with the technical aspects of using the hardware (probes, meter) and software (platform, calibrating and graphing functions) and the cognitive aspects of the technical use, for example why the need to calibrate, the interpretation of data displayed including the less explicit but implied meanings of data that are displayed on the screen. In order to reduce the cognitive load when using a more complex type of ICT resource, the student needs to develop some initial literacy with its use. This means being provided with the necessary time to explore the equipment with small tasks embedded into the learning in order to develop the necessary skills and understanding related to its use. The time taken to explore and learn well the different functions of the technology is worthwhile because when these skills and knowledge become so familiar the students will be able to use the technology intuitively and intelligently to carry out the data logging task with more success and less frustration. The argument would be similar for using other sophisticated software such as *Excel* and the *Crocodile* series, where the student would need to be familiar with the range of features associated with each of these pieces of software in order to work with it effectively. Familiarising means that the knowledge is stored in the long term memory and is called upon by the working memory when it is required to support the learning undertaken.

Framework for Empowering the Development of Scientific Literacy Through Digital Literacy and Multiliteracies

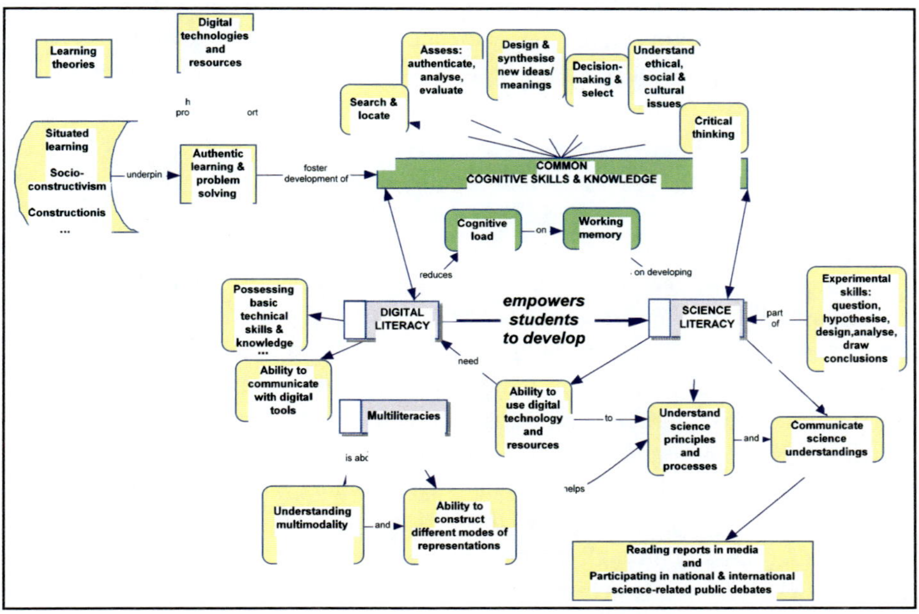

Figure 4. Conceptual framework for the empowering of scientific literacy development through digital literacy and multiliteracies (modified from Ng, 2010).

This chapter has put forward an argument that students could be empowered to develop scientific literacy if equipped with good digital literacy skills and knowledge. The points argued in the chapter are summarised in the framework shown in Figure 4. The framework

- broadly defines the capabilities of digital and scientific literacies and captures the skills required to develop them
- shows the relationships between the two literacies in that
  - (i) there is a set of skills that are common to the development of both literacies. These are critical thinking and research-based skills that have been discussed at the beginning of the chapter and
  - (ii) the ability to use digital technology and resources to better learn science content as well as communicate understandings is through multimodality
- shows that digital technologies have the capacity to provide for and support authentic learning opportunities to foster the development of both digital and scientific literacies.
- shows that digital literacy proficiency would reduce cognitive load when students are learning science using ICT. This would increase the working memory capacity to focus on the development of scientific literacy.

CONCLUSION

Digital technologies offer many advantages to students in their development of scientific literacy. Students with proficient digital literacy know how to use digital technologies effectively to search for and evaluate information to fill in gaps that they may have in the learning of a topic. They know how to use digital resources to construct content and ideas to show how well they have understood concept(s) learned. They develop their multiliteracies skills (as part of digital literacy) and knowledge of multimodality to learn science through single and/or multiple modes of representations and through multiple representations.

Using ICT for teaching and learning science requires practise and opportunities to familiarise. The more use of ICT, the more familiar science teachers and students will be with the functions, applications, issues and responsibilities associated with its use. When familiarity occurs and the knowledge and skills gained from this familiarity are stored in the long-term memory, the use of ICT becomes intuitive and automatised. When this happens, the focus of the learning will be on the science tasks at hand and the learning attention will not be distracted by the technological difficulties, causing less frustrations and achieving more success with the learning.

Chapter 6

Multimodal-Supported Science Learning: Examples of Pedagogy

Introduction

In science education, the aims of using ICT are to support the learning of science concepts - through games and simulations; obtaining authentic science-related experiences such as the space projects at NASA website or virtual museum tours; enabling data collection through data logging equipment; communicating and discussing results and ideas through graphs, spreadsheets, email and possibly social networking communities, and synthesising and creating knowledge products such as a science digital story, blog or a wiki. As discussed in previous chapters, ICT-enabled affordances that support pedagogies in science education include a range of hardware and software. Examples of these are the interactive whiteboard, digital microscopes, simulations, interactive worksheets, electronic laboratory notebook, science websites, audio and video recording devices such as flipcams, multimedia editing tools, Web 2.0 technologies and data logging equipment and software. The ability to use these technologies for effective teaching and learning would require a reasonably level of digital literacy from both the teacher and the students. Teachers with an understanding that multimodality is a characteristic of scientific texts and that ICT enhances this aspect of the text (Knain, 2006) would be better prepared in their teaching to develop their students' scientific literacy. Knain's multimodality assertion is echoed by Prain and Waldrip (2010, p. 1-2) who stated that:

> ...learning the particular literacies of science is crucial to developing science literacy. These literacies include all the signifying language practices of science discourse, including verbal, visual, and mathematical languages, as well as understanding the purposes and rationale for these literacies in representing scientific reasoning, practices and processes. Knowing how, why, and when to integrate the use of tables, graphs, diagrams and written text is crucial to representing scientific processes and claims. In this sense, every representation in science makes claims about the natural world, and learning science entails understanding the bases of these claims as a form of knowledge production. This capacity to construct and interpret science texts, what Norris and Phillips (2003) consider fundamental to demonstrating science literacy, therefore depends on students developing both procedural and conceptual knowledge entailed in scientific representation.

Hence, multimodality and multiple representations emphasise the many ways and contexts in which we communicate and develop understanding of science. This chapter will describe examples of pedagogy that uses ICT-enabled multimodality in developing science literacy. It will also discuss learning theories that underpins the development of science knowledge and skills in ICT-enhanced learning environments.

ICT-Enabled Multimodal Means of Developing Scientific Literacy

Some of the ways students learn science, via the different modes of representations enabled by ICT, are (see also Chapter 3, Table 3 and Chapter 4, Table 2):

- through audio means, for example listening to podcasts for instructions/lectures or voice recording explanations/narratives, allowing students to learn at their own pace
- from ebooks, websites and interactive CDs that accompany the static-based text books where modes of representations are limited to text and images
- through recording experimental work by taking digital images or video recordings of procedures and results that could be incorporated into their reports. They use *Excel* to plot graphs or create tables and drawing software to construct diagrams

- by sharing their experimental data from practical work with a wider audience through the posting of the images taken or video recordings of experiments on *YouTube*, *VoiceThread* or learning management systems. Others may share their experiences through descriptions on blogs and inviting peers to comments
- through the numerous science software packages that assist students to learn through drill and practice or visualisation of abstract concepts
- through the huge amount of information that is available for almost every science topic on the Web. This and other Web-based resources such as animations, videos and interactive simulations enable students to take their learning to the depth that they desire
- through online science learning communities such as listserves, virtual conferences and open access encyclopedia such as *Wikipedia*

Whether students learn science individually or collaboratively, or through listening to podcasts, reading electronic text books or accessing information on the Web, it would require them to be able to evaluate the information, identify key concepts and their meanings and make appropriate connections between concepts. The student may be required to conduct further research if there are gaps in the understanding of concepts under study. Students undertaking further research could obtain information from their teacher, consult reference materials in books or online and post questions in online learning sites. In approaching online communities and forums to help clarify the student's thinking about a problem, (s)he would search for appropriate websites where scientific Q & A forums are available. For higher education students, it would be appropriate to visit and post questions at the *Ask the Expert at Scientific American* website[1]. For school students, similar sites to post questions are the *Ask Science Questions* website[2], the *Science Club*[3] website or the Science and Mathematics section of the *Yahoo* site[4]. Understanding the 'culture' of virtual learning communities such as these and other virtual forums, and having the necessary academic and social skills to communicate in these environments are important aspects of developing students' scientific literacy. These skills are embedded within the three dimensions of digital literacy (see Chapter 3)

[1] (http://www.sciam.com/askexpert_directory.cfm
[2] http://www.asksciencequestions.com/
[3] http://scienceclub.org/kidquest.html
[4] http://uk.answers.yahoo.com/dir/

and digitally literate students will be able to participate successfully in these virtual environments.

Demonstrating and Communicating Science Understanding through ICT

As discussed in Chapter 2, one aspect of being scientifically literate is the ability to demonstrate understanding of scientific principles and processes learned. Another aspect is the ability to question the integrity of evidence shown and any ethical or social concerns that may be associated with it.

Digital tools provide effective means of constructing science understanding and communicating it. Digitally literate students who are well-versed with an understanding of the nature and operation of the different types of available software can selectively use these tools to demonstrate science understanding and use them for the synthesis of new ideas. At the operational level, it is essential that students have mastery of the features that are offered by a particular piece of software and the technical skills to operate them. For example, *Word* has numerous features that could be used to construct interactive worksheets that are multimodal. An example of creating an interactive worksheet using *Word* on the water cycle for primary students would include:

- the use of the cyclical shape in *SmartArt*
- text boxes for writing text and for moving images around the document
- the use of background colours to highlight differences
- inserting objects (images e.g. clouds; sound e.g. narration and video files e.g. process of evaporation) and
- the use of hyperlinks.

In creating presentations to convey understanding, the student should be able to decide if a *Prezi*, *PowerPoint* presentation or a Web page is the better way to get across the ideas that (s)he would like to convey. For example, if the student wanted to show the number of different food chains in a food web, using animated food chains where one chain at a time shows up on the screen would be a clearer way of presenting the information. This could be achieved using *PowerPoint* or *Flash* animations. Presenting the student's understanding in this manner would demonstrate clarity of thinking in his/her mind of what food chains are and how they are connected to form a food web. In other *PowerPoint* presentations where there is a need to show a big table that cannot fit into a *PowerPoint* slide, a hyperlink from the *PowerPoint* presentation to a html page that shows the table would be a better way of presenting it. In a web-based page, the student could scroll up and down the page to show a complete table to the audience.

Developing Scientific Literacy through Collaborative Work in Online Environments

Developing skills to work collaboratively in teams to produce meaningful outcomes is an aspect of being scientifically literate. Collaborating through WebQuests projects and the use of Web 2.0 technologies as collaborative and display tools are ways of fostering the collaborative dimension of this literacy.

Webquests

WebQuests are inquiry oriented, web-based activities requiring students to work in small teams to explore online a large body of information in a content area and complete a research task (Dodge 2007). WebQuests are created by teachers according to a standard format (Dodge, 2007) and are usually uploaded to the school's intranet or server. They could also be uploaded onto sites such as *Wikispaces* and *GoogleSites*. A WebQuest poses an authentic problem and provides relevant websites as resources to direct students' learning. The six components in a standard WebQuest design (Dodge, 2007) are: (i) *Introduction* where background information is provided (ii) *Task* where the finished product that is expected from students is described (iii) *Resources* where primarily web-based resources are listed (iv) *Process* where the WebQuest process is broken up into steps providing a framework for students to follow (v) *Evaluation* where the criteria for the evaluation of the product, usually in the form of an evaluation rubric is outlined and (vi) *Conclusion* which brings closure to the quest and students reflect on what they have learned and propose future directions where appropriate.

WebQuests are often multidisciplinary in its approach and promotes collaborative learning where authentic, real-world problems have to be solved. A well-written WebQuest requires students to analyse a variety of resources and use their creativity and critical-thinking skills to derive reasonable solutions to the problem (Yoder, 1999). For example, in a group work task in search of the perfect athlete, students take on different roles such as an anatomist, a nutritionist, trainer and statistician to investigate the various perspectives of creating a strong performing athlete. Solving problems such as this relates closely to students' daily experiences (Raizen, Sellwood, Todd, & Vickers, 1995) and fosters higher order thinking skills when they have to design, gather information, analyse, evaluate, synthesise and communicate the solution in solving the problem given in the task. Students who are skilled technically in navigating and undertaking research on the Internet would be able to focus their energy on the cognitive aspects of solving the problem without overloading the working memory with technical problems.

Collaborative Learning Using Web 2.0 Technologies

The Web has traditionally been a source of information for learners. In recent years, hosted services have been developed - Web 2.0 technologies that allows for interactions online. These technologies include building wikis and/or blogs collaboratively, sharing videos and slides and learning socially in learning management systems or social-networking sites. Web 2.0 is also known as read/write web where people are both consumers and creators of information and where the information could be their online presence, written comments, tagging of objects, a remix of content, or the construction of original content (Bell, 2011). These technologies are enablers of science learning and the development of scientific liteacry.

Learning Theories That Underpin the Development of Scientific Literacy in Digitally Enhance Learning Environment

Social-Constructivism

The learning theories that support the development of scientific literacy in digitally enhanced environments are educational constructivism and constructionism. Educational constructivism has been highly influential in Western science and mathematics education in the last few decades. It draws on Piaget (1955, 1972) and Bruner's (1960, 1966) cognitive and Vygotsky's (1962, 1978) social learning theories. It posits that the learner is an active participant in the construction of his/her own knowledge and that prior knowledge and a socially interactive environment influence this learning. The underlying principle in cognitive constructivism is that knowledge resides in individuals and cannot be given or transmitted whole to learners by their teachers. Learners must construct their own knowledge in their minds and progressively build their knowledge through experiences. Real learning can only take place when the learner is actively engaged in the process, either at the operational level where the learner is engaged in physical manipulations and/or at the cognitive level where he/she is mentally processing incoming information and stimuli. Learning science in a technologically mediated environment occurs when the learner interact with the often non-linear material that is displayed on the screen. (S)he self-directs his/her own learning by actively analysing, evaluating, making decisions and creating while manipulating the material at hand. (S)he will constantly have to compare his/her own prior knowledge of science with that presented in the learning environment and seek means of re-confirming this prior knowledge or to de-construct previous knowledge and re-construct new ones. The role of the teacher in a constructivist learning environment is to scaffold the learning and to provide the students with opportunities to be actively engaged in their own learning by:

- eliciting students' existing views on a concept
- using these views to target teaching
- providing the learners with experiences to test their ideas in order to assist them with deconstructing and reconstructing their prior views
- ensuring that teaching is student-centred where opportunities for exploration, discussions, working in groups and problem solving are made available to them. 'Hands-on' approaches are advocated in constructivist-based learning.

Vygotsky's (1962, 1978) social constructivism makes similar assertions to the cognitive aspect of constructivism about the active engagement of individuals in their learning but his theory places more emphasis on the social context of learning and in particular on the role of 'mediating agents' such as the teachers. In his theory, the learning process involves interaction with and through other individuals where culture and society will influence the learning. A difference between cognitive and social constructivism is that in the former, the teacher plays a limited role whereas in the latter, the role of the teacher is active and involved in helping students to grasp concepts by guiding and encouraging engagement in activities such as group work. Central to Vygotsky's arguments is the role of others (for example peers and parents) in mediating the learner's access to new experiences and knowledge. The social aspect of constructing knowledge is particularly significant in a technological society where young people socialise frequently in networked online communities and through informal learning with peers and family members that takes place outside of school time.

Connectivism

In recent years, Siemens (2005) proposed the theory of connectivism for learning via technology in a digital age. Connectivism describes the effect of technology on how people connect and communicate digitally in networked communities. The key elements of connectivism as summarised by Boitshwarelo (2010) are (i) learners connected to a learning community benefit from it while also contributing information to it (ii) the community is viewed as a node and is part of a wider network of nodes (iii) due to the constantly changing nature of knowledge, it needs to be continuously evaluated for validity and accuracy (iv) knowledge is distributed across an information network and across multiple individuals and (v) knowledge is connected in an interdisciplinary manner and in its construction process in the Internet environment. Examples of connectivism being applied can be seen in online communities participating in building blogs, glogs, concept maps and wikis (such as the example of the WebQuest described above). Research into the theory of connectivism is very new still and empirical evidence to support it is not available yet.

Constructionism

Blended into constructivist learning theory is the theory of constructionism. In his books *Mindstorms* (1980) and *The Children's Machine: Rethinking School in the Age of the Computer* (1993), Papert linked constructivism to technology. According to Papert, students are more motivated and engaged in learning when constructing a public artefact that others will see, critique and use. The artefact could be a sand castle, a program for a game, a *PowerPoint* or a theory of the universe. An example of constructionism in the development of science literacy in a digital environment is the production of a wiki, a glog or a video on a science topic. Team collaboration fosters social interactions, with each team member bringing with him/her his/her prior knowledge on the topic that is shared in the construction of the artefact. At the level of the individual team member, (s)he is actively constructing new meanings through (i) active research and assessing of new information (ii) contributing and responding to discussions from other team members and (iii) selecting relevant bits to build the artefact.

EXAMPLES OF PEDAGOGY OF MULTIMODAL-SUPPORTED SCIENCE LEARNING

In the following sections, I have selected a few examples to illustrate the use of ICT-enabled multimodal means of teaching and learning science. Learning in these examples is based on the social-constructivist and constructionism theories of learning.

Example 1. Multimodal Representations with Handheld Computers

This case study is based on a recent research study (Ng & Nicholas, 2011) that was conducted with year 7 students in a school with a pocket PC (PDA) program. The presentation of this case study in this chapter will focus on the multimodal representations that students used to demonstrate their understanding while studying about the circulatory system. The students' representations of understanding were captured with a video-capture software that was installed in a set of PDAs that eight of the volunteer students used. The PDAs the students in the school used were HP iPAQ hx2490b handheld computers that ran on Microsoft platform. The activities associated with learning about some aspects of the circulatory system using the PDAs are shown in Appendix I. One of the activities required the students to determine their own average resting- and after-exercise pulse rates and compare them with the averages of the rest of the class. This case study will discuss the different modes of representations that were constructed by the students in addressing the following tasks:

(i) determine and compare the averages of their own individual pulse rates before (resting) and after exercises
(ii) determine and compare the average class pulse rates for 'resting' and 'after exercise' with their own individual pulse rates and
(iii) construct graphs or histograms for the number of students with pulse rates in each of these categories
71-80; 81-90; 91-100; 101-110; 111-120; 121-130

The video-capture software recorded the screen-based actions and conversations that took place between the students and their peers while working on the problems. The recorded materials, called 'videosnags', were analysed for five of the students as they worked on the science- and mathematics- integrated task. There were some technical difficulties with the other three PDAs during the activity and the actions were not correctly captured.

Pulse Rate Data Collection and Finding Averages

The students recorded their individual pulse rate data in either *Word Mobile* or *Excel Mobile*. The two students who used *Word Mobile* either scribed or used the keyboard to enter the data. Three of the students entered their individual and class members' pulse rate data directly into *Excel Mobile*.

The videosnags showed that the year 7 students used Bluetooth to collect the pulse rates of a few of their classmates' pulse rates but observations showed that they were frustrated with the slowness of the process, prompting them to move around the class with their mobile devices to obtain the pulse rates of their peers and entering them directly into their PDAs.

The videosnags from the PDAs showed that the students used three ways of obtaining averages of individual and whole class' pulse rates.

(i) By using the PDA's calculator (this is for the students who entered their data in *Word Mobile*)

For students who entered their data in *Excel Mobile*, the two ways of calculating the class averages were:

(ii) by entering in the formula bar, the formula of the sum of the class pulse rates divided by the number of students
(iii) by using the 'average' function in the *Excel Mobile* spreadsheet as shown in Figure 1. The figure shows the student highlighting

(selecting) the 'average' function (Figure 1a). It shows the average of the pulse rates as 70.833 in the bottom bar of Figure 1b.

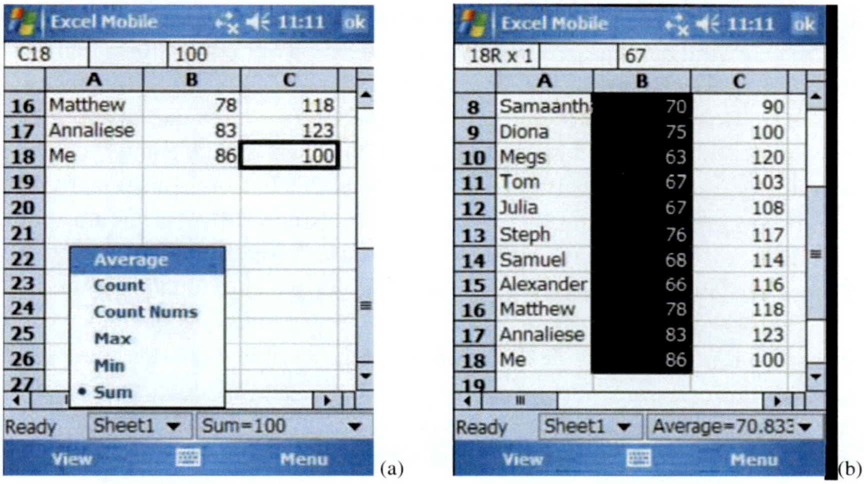

Figure 1. Using the 'average' function in Excel to find the class average pulse rate.

The videosnags provided insights into the students' knowledge of spreadsheets. The two students who manually calculated the averages do not appear to be confident users of the software and even though they have been taught to use it before, had chosen to use something they were more comfortable with – manual calculations with the calculator. For the other three students, their level of digital literacy, both technically and conceptually, with *Excel Mobile* was sufficiently high to use the software in the correct ways.

Modes of Representation in Comparing Pulse Rates

The three students using *Excel Mobile* constructed graphs to compare the averages of the class' pulse rates at rest and after exercising. Three different ways of representing this was demonstrated. Figure 2(a) shows that all the 'resting' pulse rates for each member of the class were plotted together on the left of the figure and all the 'after exercise' pulse rates on the right. The students also carefully chose the colour scheme (shown in a key at the bottom of the chart) for representing each student's bar graphs so that a comparison could be made by matching the colour graphs on the left with that on the right. The second student, however, represented the comparison by plotting each student's before- and after-exercising bar graphs next to each other (Figure

2b). This would be a clearer way of showing comparative data for an individual student. The third student constructed two charts – one for before-exercising and another for after-exercising. Figure 2 (c) shows the chart before-exercising. Only student 2 undertook the third task of constructing a graph to show the number of students with pulse rates in each of the categories of 71-80, 81-90, 91-100 etc. The student had chosen to represent this as a pie chart. While the percentages are shown in the chart, their relationship to the categories is not shown. A key is required for the chart to be meaningful.

The videosnags showed the students' competency, both technically and conceptually in constructing these graphs. The manner of presentation of the data indicates different interpretations of the task. Visually, the representation in Figure 2(b) makes it easier for viewers to compare individual's before- and after-exercise pulse rates. While these final representations were saved in *Notes* and did not require video-capture software to view them, the videosnags identified, through transitional representations, whether a students' inability to produce graphs on *Excel Mobile* was due to a lack of skills/knowledge of *Excel Mobile* or a lack of understanding of the task undertaken leading to the inability to choose the appropriate modes of representations for the activity. The students who used *Word Mobile* to represent their understanding of the tasks did not get to the graphing stage as this was not possible to do without data in a spreadsheet. This demonstrates the importance of selecting the most appropriate software to match the task in order to achieve the desired outcomes.

The video-capture software was also able to capture audio representations of understanding. For example, the understanding of one of the students was captured through a conversation where she had asked her partner "how can you have one average for the whole class?". This student appeared to know the mechanics of finding averages but lacked the conceptual understanding of what 'average' meant. This case study reinforce the argument for this book that being digitally literate would enable a student to demonstrate understanding through constructed modes of representations with the most appropriate software. In this case, the literacy is associated with the use of the *Excel Mobile* software and understanding the similarities and limitations of the spreadsheet in a mobile device compared to a desktop-based *Excel.*

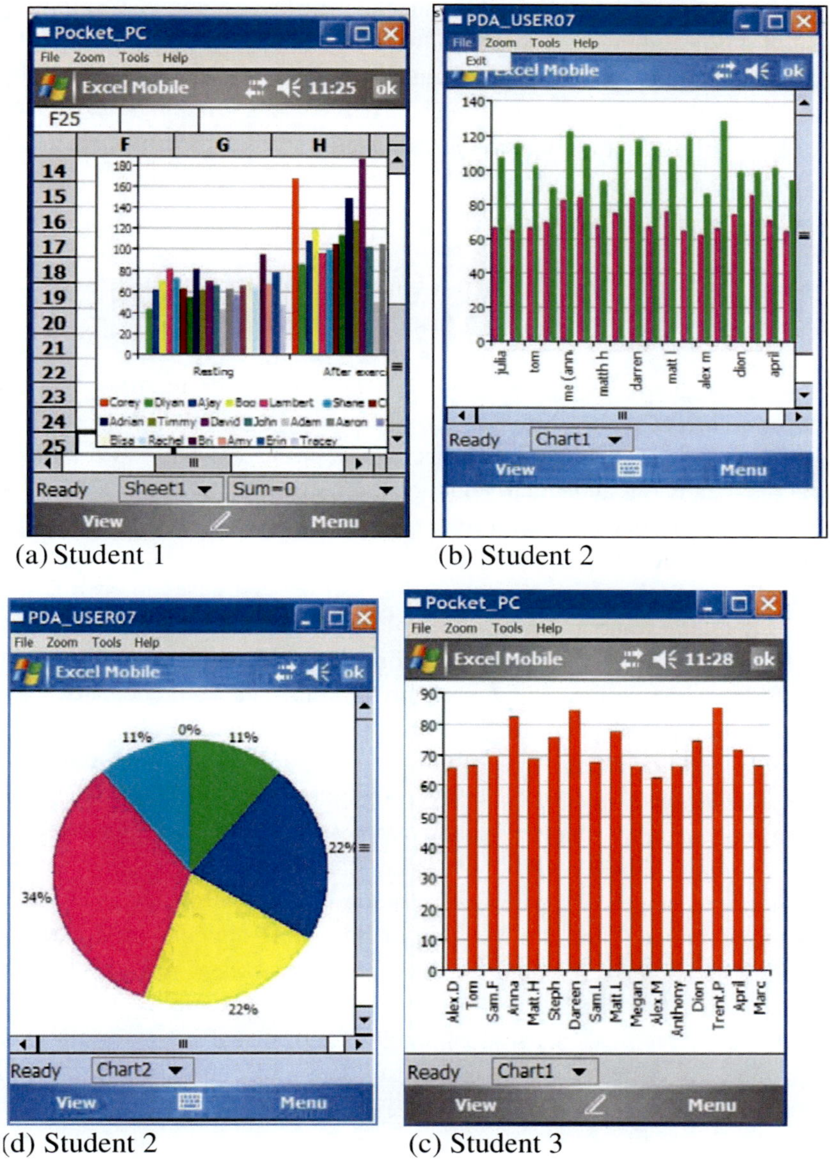

(a) Student 1 (b) Student 2

(d) Student 2 (c) Student 3

Figure 2. Figure (a)-(c) shows screen shots of the different ways of graphing in order to compare individual students' pulse rates before- and after-exercise. Figure (d) shows the screen shot of student 2's graphing of the different categories of the class' pulse rates.

There are specific technical and conceptual skills and knowledge that are required to use spreadsheet applications and it is crucial for teachers to dedicate some time in helping students to develop them systematically, especially for the weaker students. The study also shows that ICT offers choices in enabling students to demonstrate their understanding using different modes of representations.

Example 2. Converting Text to Visual by Mapping Online Collaborative Dialogues into Concept Maps

Web 2.0 technologies such as wikis and blogs offer multiple ways of working collaboratively online. In educational institutions, learning management systems (LMS) such as *Blackboard*, *Moodle* and *Edmodo* provide the means for students to collaborate on projects virtually in their own time and space. In discussion forums, tasks or topics are set for students to participate as whole class activities or in small virtual teams. Depending on the number of strands started in a conversation and the responses in each strand, dialogues that take place in an online learning environment are often disjointed or patchy as responses to some discussion items are numerous and lengthy while others are skipped over or completely left out. This could be caused by a number of simultaneous strands of conversation that students either respond to or ignore depending on their level of knowledge and interest. The discontinued or disregarded threads of communication combined with the delay and overlap in various concurrently running strands may result in patchy and unsystematic student learning. Consequently, recorded script in learning management systems often appears somewhat random with unrelated pieces of conversations, thus raising the question of how well students are connecting the learning. In addition, it would also be difficult to assess how much the non-contributors to the discussion are learning. This example explores the consolidation of textual discussions online into visual representations as concept maps. Specifically, students consolidate and internalize their newly gained knowledge, which has been generated and made explicit during the online discussions. By getting students to transform their learning from text to picture, it takes their learning to a higher level of thinking (Bloom, 1956; Anderson and Krathwohl, 2001). The continued fostering of these skills is essential in developing students into critical thinkers and problem solvers.

The process of getting students to transform text to picture would require the student to identify key concepts - both present and absent in the recorded dialogues and to then show their interrelationships in a concept map. This means engaging students in multimodal ways of representations, from linguistic to visual representation in a map. If voice-recorded and multimedia files are inserted into the concept map, audio and multimodal representations are also used. In this type of task, the student will be required to reflect, analyse and undertake further research in order to organize the information from the online dialogues into a structured system (concept map) to demonstrate their understanding of the topic. The theoretical frameworks of constructivist and constructionist learning would underpin this type of individual learning.

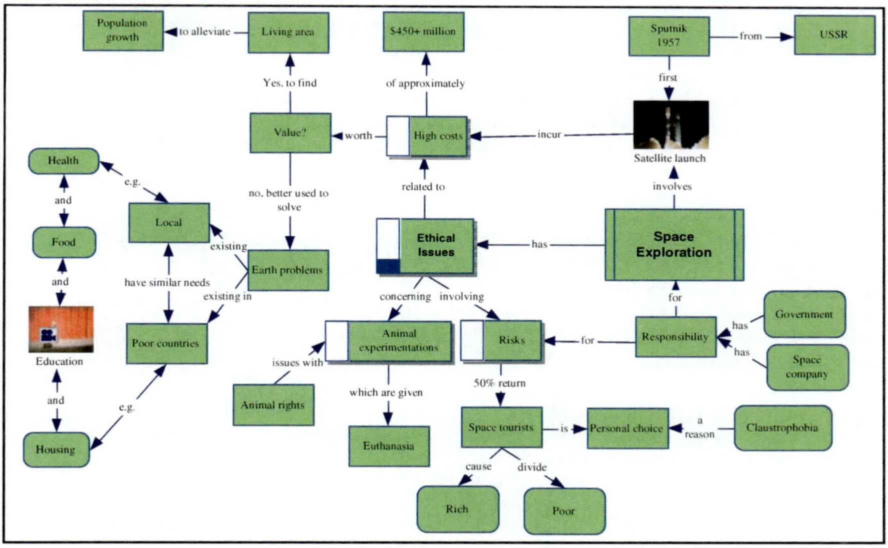

Figure 3. Concept map for Space Exploration resulting from pre-service teachers' unedited online collaborative learning (Ng, 2010, p.23).

Two examples of the transformation of online dialogues to visual representation using the concept mapping software, *Inspiration* are shown. The dialogues are taken from two separate online learning activities. The first was conducted with a group of pre-service teachers studying the unit *Science and Technology in Contemporary Society*. The focus of the online discussion task was about ethics where students worked in small teams of 3-4 students to decide on a topic of interest related to science and/or technology and discuss the ethical perspectives of the topic.

Table 1. Pre-service teachers' online dialogue on Space Exploration and identification of key concepts for concept mapping

Space exploration (selected dialogues, unedited)	**Key concepts identified**
STU3>>agree, coz i wanna go to space STU2>>can't afford it at the moment though STU3>>so unless i win lottery STU3>>In terms of ethical debates and issues, the arguments underlying space exploration is something that is not often raised in the media, such as human cloning and abortion for example….Should so much money be spent on space exploration with so many problems on our own planet?	Poor Rich Value of space exploration Problems on earth
STU1>>hehe..i couldnt go to space...i'm scared of being confined in small spaces	Claustrophobia
STU1>>i liked how you included some history STU2, ... (cut and pasted from short essay STU2 produced) The ethical debates surrounding space exploration started at the dawn of the 'space race' in the late 1950?s when the former USSR successfully launched Sputnik, a satellite into the outer atmosphere of earth. From that instant, governments and organizations have hurled things into space where the moral and ethical concerns have been intensively debated. The initial space flights sparked debates about animal rights where numerous animals such as dogs and primates were sent into space to predict how humans would respond to this new and strange environment. These animals however never returned to earth but were euthanized near the end of their 'missions'.	History of space travel 1950 USSR Sputnik Moral concerns Ethical concerns Animal experimentation Animal rights
STU2>>cool, so then move on with... STU1 (continues cutting and pasting from her own work>>The development of space exploration has led to the possibility of tourists in space. As a result the ethical considerations of such a possibility need to be considered. The probability of not returning from a space expedition (to the moon) is 50%. Is it ethically viable to allow tourists to enter into space with such a high probability of not retuning? Would it be up to the company running the space expedition, the government or the tourist to decide whether they were going to accept this risk? Would the desire for money of the company governing the space tours outweigh their desire to make tourists fully aware of the potential hazards of such an expedition? Additionally, the cost of being a tourist on a space expedition is undoubtedly going to be rather expensive as launching a space shuttle costs approximately $450 million. Thus, space tourism will only be available to those who are quite affluent, consequently creating yet a further divide between the rich and the poor. So, is it ethically just that only the rich have access to, and the opportunity to obtain all of the benefits that might arise from, tourism in space?	Space tourists Risks 50% return chance Company greed Government responsibility Personal choice Costs 450 million launch space shuttle Gap widens between rich and poor
STU1 (cut and paste continues) >>The risk to, and loss of human life is just the tip of the iceberg when it comes to the moralities surrounding space exploration. Such topics include unnecessarily excessive funds being diverted to space programs in many countries when that money could be better spent on healthcare, food and housing for the poor or education. Other people are concerned about such things as storage of nuclear waste in space, isn?t it enough that we fill our own planet to the brim with highly toxic rubbish that we no have to resort to shooting it into space?	Money better spent Poor countries Healthcare, food, housing, education Nuclear waste in space
STU3>>The possibilities that space exploration leads to could be enomous, however the ethics associated, and fundementally those to do with economics, could have a colossal influence on the extent to which such exploration progresses.	Economic ethics

They then wrote, collectively as a group, a short essay on it. The dialogues are real but the concept map generated from them is simulated, as an example of how educators could take an online discussion task and extend it into a higher order thinking task where students will need to (i) read the conversations again (ii) identify key concepts that have been discussed (iii) identify missing concepts and (iv) construct a concept map using the key concepts identified and place linking words between them. Table 1 shows selected dialogues where key concepts were identified and Figure 3 shows the resulting concept map created. The topic that this group of students chose to focus on was Space Exploration. The students had done some prior research and engaged in asynchronous discussion before moving to this synchronous discussion in the chat room, from where the dialogues in Table 1 is obtained. The conversations in the table is presented unedited. Included in the concept map are an image of a satellite launch and a video on educating about space exploration. More information on the image and the playing of the video is achieved through hyperlinking. Being skilled with branching literacy could enable the student to add more information to the concept map via hyperlinking in this way. The concept map is multimodal with text, image, video and other visuals such as linking arrows and different shaped boxes (for keywords) where meanings are also assigned to.

Similar to the first example, the second example is based on an online discussion (Table 2) between Year 9 students debating about solar vs nuclear energy. The students had attended a summer camp on the theme of *Sun, Science & Society* and their learning was extended after the camp for a further six months online. One of their online tasks was to debate about these two sources of energy. Figure 4 shows a simulated concept map from the students' online conversations.

Table 2. Year 9 students' online dialogue on nuclear energy and identification of key concepts for concept mapping

Nuclear Energy (selected dialogues, unedited)	**Key concepts identified**		
	General info	**Pro**	**Con**
GH:nuclear fission is very "dirty" coz it has alot of radioactive waste...that is a fact and cannot be denied...but nuclear fussion releases HELIUM which is extremely safe...I tink in de next couple of years once nuclear fusion is possible (close enough from the Z-machine), it'll be really safe...since nuclear fusion nids a lot of energy to get it started, all they nid to do is to get rid of de input energy n de reaction would stop as there wouldn't b anymore sufficient energy. dat is completely safe...	Fusion – less radioactive waste; releases helium, safer; technology unavailable	Fusion – safer	Start up energy: Fusion - huge Fission – less Fission releases radioactive waste

Table 2. Continued

Nuclear Energy (selected dialogues, unedited)	**Key concepts identified**		
	General info	**Pro**	**Con**
PP: to add on, the fuel most widely used for nuclear fission is urainium. it's non-renewable and it's found in rocks all over the world. nuclear plants use U-235 because it's atoms are easily split apart. Even though U-235 is more common than silver, it's quite rare......during nuclear fission, the neutron hits the urainium atom and makes it split, releasing a great amount of energy as heat and radiation	uranium 235 for fission; neutron splits uranium atom		non-renewable
AL: that radioactive waste matter which you are so gung-ho about has its own uses.The radioactive rays produced by this waste matter is used to sterilize medial equipment like syringes.Gamma rays are used to kill germs inside the SEALED packet which holds the syringe. These rays are also used in chemotherapy to destroy the cancerous growths, and can detect parts of the body which aren't functioning properly. FURTHERMORE,radioactive waste can help scientists measure things like ocean current flow by mixing a radioactive liquid with the seawater, and a Geiger counter (a machine which measures the distance of a radioative substance) can then give a reading (the liquid loses its radioactivity after a short time). Sources:Childcraft (A world book encyclopedia publishing)		Radioactive waste uses: Sterilisation Chemotherapy Science investigations	
CH: the radiation going on inside the core produces a massive amount of heat, which heats water which turns a turbine. Yep; Nuclear energy is just a fancier version of coal energy. But, the waste from a nuclear plant is so powerful, that it would kill the person along with the cancer, or even contaminate the very ocean/sea/lake they wanted to test.		Heat generate electricity	Cause cancer
AL: Nuclear energy is mostly derived from uranium-an element which can be found on the earth's surface (it needs to be mined).Nuclear energy is VERY clean.It does not emit any carbon dioxide or any harmful gases into the atmosphere.This is very convenient-so convenient that 20% of USA's energy is nuclear,and 80% of France is nuclear. Another thing about nuclear energy is that it doesn't depend on the weather/climate.If there is no sun,there wouldnt be any energy!	No carbon dioxide or harmful gases	Clean energy Non-sun or weather dependent	
CH: Ahhh yesss.... when you say how 'clean' this is, you conveniently forgot to mention the tons of radioactive material leftover, that gradually builds up and has to be stored in cement vaults....'Chernobyll', as AL helpfully mentioned, is the most famous case of a nuclear power station blowing up in a massive release of atomic energy and eradicating everything in a 5 mile radius, or contaminating anything in neighbouring countries with radioactivity. Not sure about the specifics, but somehow it went out of control, and exploded, contaminating everything with radiation, as far away as Italy.Lots of cement had to be dumped on what remained of the plant, and if you took the cement off, it would still be radioactive today and you (And everyone in a 10 mile radius) would develop a nuclear tan and die within the hour. Real safe, eh? Nuclear powerplants take many people to operate them... scientists say that it is biodegradable after a period of time."Well, technically, this is true.If you count 'a period of time' to be about 500 million years. And that's just how long it takes to lose its radioactivity and turn to lead; the biodegradability of lead is another thing altogether....	Radioactive wastes take millions of years to lose their radioactivity	Create jobs	Accidents Contamination Biodegrade millions of years

Both these examples show that the use of multiple representations to demonstrate understanding fosters the development of scientific literacy and higher order thinking skills. If there are no online conversations available to undertake the activity above, mapping key concepts based on an essay or textbook-based information is another way to develop scientific literacy through concept mapping. For example, in Appendix II, the essay that was constructed by the group of students discussing about Space Exploration could be used by another team to construct a visual representation, using the method above.

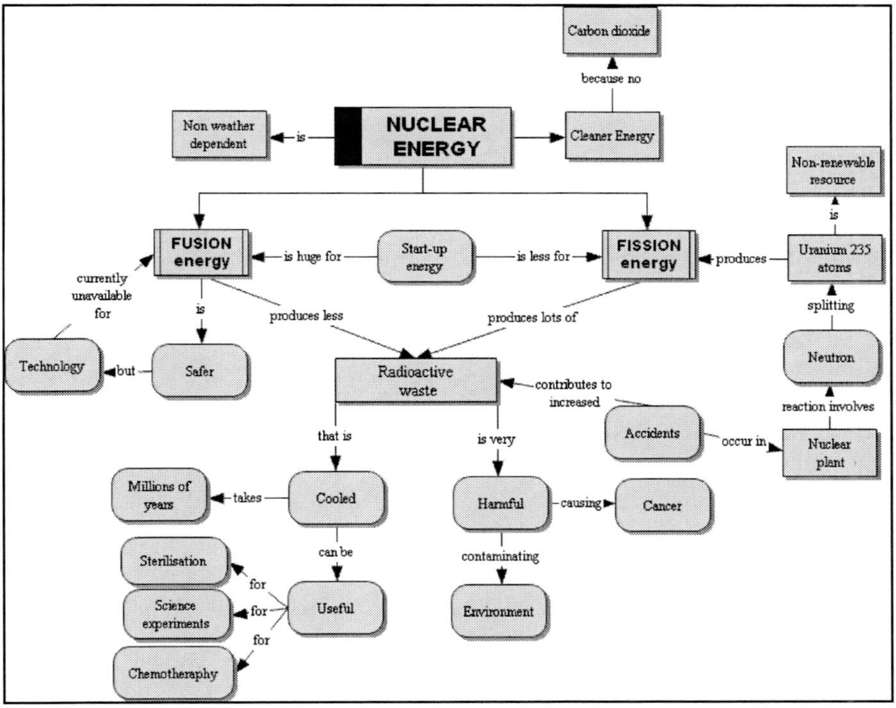

Figure 4. Concept map for Nuclear Energy resulting from Year 9 students' online collaborative learning.

The converse, that is converting a visual (concept map) into written or verbal text, could also apply as a pedagogy for developing scientific literacy through multimodal representations. Getting students to critically analyse visual images and transforming them into a descriptive product that is either word processed or audio recorded (using mobile phone, iPod, Smartpen or other audio recording devices) not only enhances scientific literacy but also visual, linguistic, audio and critical literacies.

Example 3. Promoting Multiliteracies and Self-Directed Learning: A Teacher's Pedagogy Using Onenote

This example describes an ICT-enriched pedagogy that Charles, a secondary school teacher, designed for his year 7 students studying the topic of Living Things. The concepts studied included plant and animal cells, viruses and bacteria, uni-and multi-cellular organisms and simple classifications. The unit of work was designed using *Microsoft OneNote* software. The software has capacities that include free-form information input and gathering, pages being able to be moved around inside the binder, annotations can be made by word processing or with a stylus/drawing tools, and multimedia materials and weblinks could be embedded. It allows for collaboration between multi-users and the user's handwritten or typed notes, drawings and audio recordings (e.g. commentaries) could be shared over the Internet with people using the *OneNote* program.

The capacities of *OneNote* provided the means for Charles, a highly digitally literate teacher, to design a unit of work that was student-centred, promoted self-directed learning and differentiated the lessons, including catering for higher ability students. It constituted a different way of learning for the students. The focus of this example is on the multiple modes of representations that were built into the unit of work that the students were exposed to in helping them develop their knowledge about living things, cells and related concepts.

The interactive, multimodal unit of work was student-led where the students undertook a variety of activities as they systematically built their knowledge and understanding of the topic. It promoted different levels of thinking skills and enhanced the digital literacy capabilities of the students. Some examples of multimodality demonstrated in the activities were:

- reviewing a video on fungi/bacteria, and asking four questions about it, sending the questions via email to another student, and when the student received another student's email, (s)he copied and pasted the questions into the working space provided on the *OneNote* page itself. The student then made an attempt to answer the questions after which (s)he shared the answers with the student who created the questions.

Students were able to collaborate (share and communicate) in online social spaces such as *Facebook* when at home. The modes of representations that students were exposed to in this collaborative activity were (i) multimodal for the video where a combination of audio, visual and linguistic modes of representations were displayed and (ii) depending on the manner the questions were answered, text, images and/or audio recorded explanations could be included – hence using single and/or multiple modes of representations for the different tasks within the activity

- working with interactive worksheets, for example, labelling a microscope on the page itself where there was a combination of visual (picture of microscope) and linguistic (text-based) modes of representations
- reading information and through answering questions, the student is synthesising understanding from the information. Depending on the questions and the means the student chose to answer them, use of images (visual) and/or text-based (linguistic) modes of representations were displayed. Alternatively, verbally inclined students could audio-record their answers in the *OneNote* program itself
- signing into the *Quizlet* website to create a set of flashcards for terms and concepts learned. Depending on whether text or images were used, it would be linguistic and/or visual modes of representations.
- draw and explain (via audio recording) simultaneously a plant and an animal cell. The students would have adopted audio, visual and spatial modes of representation here.

There were also research projects, for example, investigating the relationships between the development of cell theory and the invention of the microscope, that required students to draw on their information literacy and critical literacy skills to search, assess and synthesise new information. A video recording of this type of project, where the student 'acted' in the video, for example demonstrating and explaining different types of microscope, would show gestural mode of representations in the way hand movements and facial expressions are used. As with most video recordings, multiple modes of representations - audio, visual and spatial (e.g. three-dimensional objects such as cells) would be involved.

The software enabled Charles to provide recorded instructions that scaffolded the learning so that the students could work at their own pace. An example of scaffolding was Charle's demonstration in *OneNote* on how to audio record and draw at the same time. According to the multimedia learning theory (Clark & Paivio, 1991; Mayer & Sims, 1994; Tuman, 2008), the use of dual-sensory modes where the visual and auditory channels are used simultaneously would help students to learn better.

In reviewing the students' drawing of cells, Charles was able to track how they have done the drawing as their audio explanations would have aligned with their drawings. Feedback could be provided as a spoken (audio recorded) or a written comment. The feedback stayed within the section of the student's drawing so that when it came to revision, the student would be able to hear again what the feedback comments were.

For the students, the multimodal nature of the unit of work was empowering their learning of science concepts, offering not only motivation (from the variety of formats) but also assistance with conceptual formation where the multiple representations in their multimodal formats supported or reinforced concepts being learned. The multimodal learning was facilitated by safe (protected) interactive websites (such as swirk.com.au, worldbookonline.com), images, animations, videos, written notes, working spaces, thinking tasks, collaborating tasks, homework sheets/tasks, worksheets, note reading and others. The software has the capacity to capture a complete record of everything that a student has done, including sites visited and the written and oral annotations made over an image or a video. It also has the capacity to link notes/websites so that when it comes to revision, students could go back to the source. The students were able to control their own learning and were able to see the relationships between the outcomes, the work that they have done and the feedback provided by the teacher.

By getting the students to use a variety of digital applications and resources in the learning, they were concurrently developing digital and scientific literacy skills and knowledge. As the students use more applications and learn about different digital learning environments, they develop better digital literacy skills and knowledge. These skills gradually become second-natured with increased familiarly and the learning would be centred on concept development without the distractions imposed by the technology.

Example 4. Cross Curriculum Perspective: Aboriginal and Torres Strait Islander Education

The Australian curriculum has identified three cross-curriculum perspectives that are to be integrated, where appropriate, into key learning areas such as science. These perspectives are outlined in *The Shape of the Australian Curriculum*[5] (p. 13) as:

- Indigenous perspectives, which will be written into the national curriculum to ensure that all young Australians have the opportunity to learn about, acknowledge and respect the history and culture of Aboriginal people and Torres Strait Islanders
- a commitment to sustainable patterns of living which will be reflected in curriculum documents
- skills, knowledge and understandings related to Asia and Australia's engagement with Asia.

This example demonstrates the use of ICT to develop an understanding of an aspect of the first dot point – Aboriginal and Torres Strait Islander education. The *Australian Curriculum Assessment and Reporting Authority* (ACARA) in developing the Australian Curriculum is committed to ensuring the need for all Australian children to "understand and acknowledge the value of Indigenous cultures and possess the knowledge, skills and understanding to contribute to, and benefit from, reconciliation between Indigenous and non-Indigenous Australians." (*Melbourne Declaration on Educational Goals for Young Australians*).

The example draws on the Year 8 National Science Curriculum content in the Earth & Space Sciences where students are expected to develop knowledge and understanding of:

> Sedimentary, igneous and metamorphic rocks contain minerals and are formed by processes that occur within the Earth over a variety of timescales

An example of a multimodal resource that teachers could draw on for teaching this aspect of the curriculum is from the *National Association of Geoscience Teachers'* (NAGT) website where the *On the Cutting Edge* geosciences project[6] is presented. The website has a collection of interactive visual materials that make use of multimodal representations to assist students to achieve the learning goals for this element of the curriculum. The systematic collection of *Flash* animations displayed is shown below:

---

[5] http://www.acara.edu.au/verve/_resources/Shape_of_the_Australian_Curriculum.pdf
[6] http://serc.carleton.edu/NAGTWorkshops/petrology/visualizations/rock_cycle.html

- *Rock Cycle, Exploring Earth*[7] provides an overview of rock formation and shows some of the most common rock-forming processes. There are embedded animations that include crystallization of magma to form igneous rock, rock erosion to create sediment, transportation of sediment, deposition of sediment to create sedimentary rock, and creation of a metamorphic rock in a subduction zone.
- *Igneous Rock Formation*[8]*, Exploring Earth* uses three separate Flash animations to demonstrate the formation of igneous rocks in different environments: a) rocks forming from a deep magma chamber where the slow cooling of magma results in large interlocking crystals; b) rocks forming from a pyroclastic flow with a combination of large and small crystals; and c) rocks with small crystals created from a fast cooling lava results.
- *Igneous Rocks Classification*[9] uses *Flash* roll over features to enable students to view hand specimens of different igneous rocks that are classified according to texture and chemical composition. Views of the more common rock forming minerals are also shown. A warning that loading times will be long alerts students to be patient while downloading the resource.
- *Clastic Sedimentary Rocks Formation, Exploring Earth*[10] is a *Flash* animation that traces the formation of sedimentary rocks from a beach environment where sea water minerals cement sand grains to form sandstone. The use of an inset to provide details of the process is part of the animation. Students can pause and rewind to learn at their own pace.
- *Metamorphic rock formation, Exploring Earth*[11] is a *Flash* animation that shows the evolution of granitic igneous rock to a metamorphic rock with a focus on changes in the alignment of amphibole, plagioclase feldspar, and quartz crystals. The use of an inset to

---

[7] http://www.classzone.com/books/earth_science/terc/content/investigations/es0602/es0602page02.cfm

[8] http://www.classzone.com/books/earth_science/terc/content/investigations/es0603/es0603page05.cfm?chapter_no=investigation

[9] http://www.wiley.com/college/strahler/0471480533/animations/ch12_animations/animation1.html

[10] http://www.classzone.com/books/earth_science/terc/content/visualizations/es0605/es0605page01.cfm?chapter_no=visualization

[11] http://www.classzone.com/books/earth_science/terc/content/visualizations/es0607/es0607page01.cfm?chapter_no=visualization

provide a close up of how pressure compresses mineral grains to form new minerals is part of the animation. Students can pause and rewind to learn at their own pace.

In addition, *YouTube* videos that incorporate audio and visual modalities to reinforce concepts learned are available to assist students who have preferences for audio messages rather than just reading text. Examples include *The rock cycle* at http://www.youtube.com/watch?v=v3yJArifULo&NR=1 and the BBC video series, an example of which is *C045 Rocks and minerals* at http://www.youtube.com/watch?v=7Mwuc_8of0&feature=related

Working with these resources provides the opportunity to connect these modes of representations with the verbal explanations of the teacher, the written descriptions in the text book and the hands-on model construction in the laboratory.

An extension of the learning of this topic could be the exploration of Aboriginal art and the paints they use. The paints are made from ochre, an earthy sedimentary rock with small particles and it is the most important material used traditionally by the Aboriginal people. As a research project, the students find out about Aboriginal art, the types of colours used, the chemistry of ochre and how the colours of the paints are derived. They put in relevant keywords in search engines to search for images and information about Aboriginal art such as the Kunwinjku art form is the x-ray style where the internal organs are shown (Read, 2000). In addition, the chemistry of Aboriginal painters using earth colours could be studied. Red, yellow, white and black natural ochre pigments provide aboriginal artists with their traditional paint colours. Through research, students should be able to come up with the key concepts of Aboriginal art that are science-related such as:

- Ochre, an earthy sedimentary rock is composed of largely yellow and/or brown iron oxides. It also contains a lot the clay mineral called kaolin.
- Kaolin is produced as a result of the chemical weathering of feldspar, a softer mineral belonging to the silicate family. The symbolic representation of this chemical weathering is a hydrolysis process as shown below (Nelson, 2010):

$$\underset{\text{Orthoclase}}{4KAlSi_3O_8} + \underset{\text{Hydrogen ion}}{4H^+} + \underset{\text{water}}{2H_2O} \longrightarrow \underset{\text{Potassium ion}}{4K^+} + \underset{\text{Kaolinite (clay mineral)}}{Al_4Si_4O_{10}(OH)_8} + \underset{\text{quartz}}{8\,SiO_2}$$

- Kaolinite $Al_4Si_4O_{10}(OH)_8$ provides the white colour for the artwork
- Iron oxide Fe_2O_3 provides the reddish colour
- Limonite or hydrated iron oxide $Fe_2O_3xH_2O$ provides the yellow colour.
- Black colour could come from charcoal or manganese dioxide MnO_2.

As part of the project, students could construct a digital story about aboriginal art, with a focus on the science aspect of it. They will show their understanding by using a variety of modes of representations – write a script (linguistic) where they narrate (audio) as voice over the text, images such as completed art work and pictures of ochre, symbols such as chemical formulae and equations, and videos showing how Aboriginal people use the 'earth paints' to paint a picture. A digital story will require skills in using film-making packages such as *MovieMaker*, *PhotoStory* or *iMovie*. The students could share their digital stories by either presenting them in class or uploading them onto *VoiceThread*, a web-based application that allows students to place artefact such as images, videos, documents, and presentations at the centre of an asynchronous conversation. Class members and the teacher would need to be invited to have conversations about the displayed artefact and to make comments that are either written or voice recorded. Hence the use of this online collaborative tool, which is also password protected is safe for students to use.

Alternatively, the students could create glog, an interactive poster. The education site for glogs, Glogster EDU, is at http://edu.glogster.com/ where numerous examples in all discipline areas are shown. Students could make use of videos, images, texts and hyperlinks to create glogs online.

CONCLUSION

The chapter has described learning theories that underpin the development of scientific literacy in ICT- enabled learning environments. The multimodal ways that information can be presented using ICT provide more flexible means of learning science for students. This is achieved through written and verbal/audio texts, visual/spatial representations such as pictures, 3-D images, illustrations, graphs or a combination of audio, visual and gestural modes of representation in multimedia objects such as animations and videos. The examples provided in this chapter highlight the importance of digital literacy in embracing the variety of ICT tools that provide the means for displaying concepts in multi-modes and multiple representations to help students develop their scientific literacy.

APPENDIX I

Investigating the Heart and Pulse Rates (Ng and Anastopoulou, 2011, P. 297 – 298)

Conduct this activity using your pocket PC only. Use resources such as stop clock in the pocket PC to assist you with completing this activity. Present data collected and answer all questions clearly on the pocket PC. You will work in pairs for task 1, 4 and 13.

1. Determine your resting pulse rate.
   Each time the heart beats, it sets up a wave of pressure which travels along the main arteries. This is called a pulse wave and if you put your finger on your skin just above the artery in your wrist or neck, you can feel it as a slight throb. Sitting down, determine your pulse rate by counting the number of times your pulse beats in one minute. Record your result. Do this twice more and find your average 'resting' pulse rate in 'beats per minute'.
2. How much work does the heart do?
   Each time the heart beats it pumps about 80 ml of blood into the arteries. Calculate how much blood your heart pumps in 1, 5 and 10 minutes.
3. How many 2 litre coca cola bottles could your heart fill in 5 minutes?
4. Determine your pulse rate after one minute of exercises.
5. Stand up and do some vigorous exercises for one minute e.g. bending exercises or go for a run. Determine your 'after exercising' pulse rate. Record it as 'beats per minute'.
6. How does the 'after exercising' pulse rate compare with the 'resting' pulse rate.
7. Use your pocket PC to obtain all the 'resting' and 'after exercising' pulse rates from the rest of the class.
8. Record all the data collected and find the average class pulse rate for 'resting' and 'after exercising'.

9. What is the range of the 'resting' and 'after exercising' pulse rates of the class?
10. How does your 'resting' and 'after exercising' pulse rates compare with the average of the class?
11. Construct a graph or histogram for the number of students with pulse rates in each of these categories
    71-80; 81-90; 91-100; 101-110; 111-120; 121-130
12. Where does all the blood that the heart pumps go to?
13. Show the connections between these 8 keywords: heart, blood vessels, veins, arteries, blush, heart attack, cholesterol, food. Choose whatever way you like, for example, you could draw a flowchart or write a short paragraph or voice record a description of how the 8 keywords are connected etc.
14. Internet search. Look up the resting heart rates of 2-3 small animals (e.g. rat) and 2-3 large animals (e.g elephant). Show the heart rate next to an image of the animal. What do you notice about heart rate and size of an animal?
15. Which animal's heart rate is closest to yours?

APPENDIX II

Space Exploration: The Ethics

In terms of ethical debates and issues, the arguments underlying space exploration are not very often raised in the media, unlike those commonly discussed, such as human cloning and abortion for example. However once the statement "Should so much money be spent on space exploration with so many problems on our own planet?" is prompted, a very heated debate between opposing sides arises.

The ethical debate surrounding space exploration started at the dawn of the 'space race' in the late 1950'swhen the former USSR successfully launched Sputnik, a satellite into the outer atmosphere of earth.[12] From that instant, governments and organisations have hurled things into space while the moral and ethical concerns have been intensively debated. The initial space flights sparked debates about animal rights where numerous animals such as

[12] www.seasky.org/spaceexp/sky5d.html

dogs and primates were sent into space to predict how humans would respond to this new and strange environment. These animals however never returned to earth but were euthanized near the end of their 'mission'.

The development of space exploration has led to the possibility of tourists in space. As a result the ethical considerations of such a possibility need to be considered. The probability of not returning from a space expedition (to the moon) is 50% [13]. Is it ethically viable to allow tourists to enter into space with such a high probability of not returning? Would it be up to the company running the space expedition, the government or the tourist to decide whether they were going to accept this risk? Would the desire for money of the company governing the space tours outweigh their desire to make tourists fully aware of the potential hazards of such an expedition? Additionally, the cost of being a tourist on a space expedition is undoubtedly going to be rather expensive as launching a space shuttle costs approximately $450 million[14]. Thus, space tourism will only be available to those who are quite affluent, consequently creating yet a further divide between the rich and the poor. So, is it ethically just that only the rich have access to, and the opportunity to obtain all of the benefits that might arise from, tourism in space?

The risk to, and loss of human life is just the tip of the iceberg when it comes to the moralities surrounding space exploration. Such topics include unnecessarily excessive funds being diverted to space programs in many countries when that money could be better spent on healthcare, food and housing for the poor or education. Other people are concerned about such things as storage of nuclear waste in space, isn't it enough that we fill our own planet to the brim with highly toxic rubbish that we now have to resort to shooting it into space?[15]

The possibilities that space exploration could unfold may be enormous, however the ethics associated, and fundamentally those to do with economics, could have a colossal influence on the extent to which such exploration progresses.

---

[13] http://portal.unesco.org/shs/en/files/8460/11223752131RationalesSpaceExplor.pdf/RationalesSpaceExplor.pdf iv

[14] http://www.nasa.gov/centers/kennedy/about/information/shuttle_faq.html#10

[15] www.engin.umich.edu

Chapter 7

IMPLICATIONS FOR TEACHERS IN INTEGRATING DIGITAL TECHNOLOGIES IN SCIENCE TEACHING AND LEARNING

INTRODUCTION

Teaching is a complex activity and the process of preparing teachers to teach is similarly complex. The demands of a society that is very much influenced by the fast-changing nature of technology create additional challenges for teachers and teacher educators. In order to both reflect and engage with these changes, it is important for educators to be aware of the changing culture and expectations of the younger generation of students, who use ICT creatively to communicate and to seek and process information for both educational and recreational purposes. The challenges for teacher educators include staying abreast of new technologies and changes associated with them while simultaneously educating pre-service teachers on how to integrate technology effectively into their teaching methods. Similar challenges are faced by practicing teachers in the need to stay abreast of new educational technologies and integrating them appropriately into their teaching. ICT has the potential to offer rich learning environments that foster flexible learning and enable students to adopt multiple perspectives when learning about complex and abstract science phenomena. In order to harness this potential effectively, teachers need to be well prepared and possess a sufficient level of digital literacy to facilitate the learning. This chapter will discuss the issues associated with ICT integration in education and the implications for teacher adopting these technologies in the science classroom.

ISSUES WITH ICT INTEGRATION IN EDUCATION

Studies have shown that the level of ICT integration and its impact in classroom curricular remain ad hoc and low (Becker 2000; Cuban 2001; Kozma 2003, Ng & Gunstone, 2003; Romeo 2006). In Australia, a 2009 DEEWR[1]-commissioned report by the Strategic ICT Advisory Service (SICTAS) titled *Hot topic: ICT in preservice teacher training*[2] indicated that:

> The evidence presented in this paper strongly points to fundamental systemic flaws in the preservice teacher education system in Australia in terms of developing teacher competence in embedding ICTs in pedagogy and practice. Unless a radical approach is taken, it is unlikely that a grab bag of individual changes will make any real difference. (p.7)

The systemic flaws included a lack of quality use of ICT in pre-service teaching courses, fragmented programs in preparing teachers to use ICT and the lack of skills and confidence of teacher educators in modeling and applying ICT in their teaching. The findings are supported by findings in the 2010 report titled *ICT professional learning: National mapping of ICT-based professional learning*[3] by Educational Services Australia (2010).

To address these issues and as part of the *Australian Digital Education Revolution*, an ICT Innovation Fund which is part of the $40 million *Digital Strategy for Teachers and School Leaders* was established by DEEWR to support the professional development of teachers and school leaders in the use of ICT. Projects funded [4] include *Leading ICT in Learning* (coordinated by Principals Australia), *Anywhere, Anytime Teacher Professional Learning* (coordinated by the New South Wales Department of Education), *ICT in Everyday Learning: Teacher Online Toolkit* (coordinated by Education Services Australia) and *Teaching Teachers for the Future* (TTF) (coordinated by Education Services Australia). In addition, an initiative targeted at students is *The Australian Government's National Secondary Schools Computer Fund* where all Years 9-12 students from government schools are equipped with a laptop each. These initiatives highlight the importance of and need to develop

[1] Department of Education, Employment and Workplace Relations

[2] http://dspace.edna.edu.au/dspace/bitstream/2150/54714/1/SICTAS_HT_pre-service.pdf

[3] http://www.deewr.gov.au/Schooling/DigitalEducationRevolution/DigitalStrategyforTeachers/Documents/ICTMappingFinalReport.pdf

[4] http://www.deewr.gov.au/Schooling/DigitalEducationRevolution/DigitalStrategyforTeachers/Pages/ICTInnovationFund.aspx

digital literacy in students and teachers for the successful integration of digital technologies into the school curriculum.

Research (Hew & Brush, 2007; Mishra & Koehler, 2006; Ng & Gunstone, 2003;Tebbutt, 2000) has also shown that two of the main barriers of effective and sustainable integration of ICT in schools are (i) a lack of ICT related skills and knowledge in teachers and (ii) the time required to develop them . The successful integration of ICT in schools depend on a range of elements, the two most important ones being (i) leadership providing vision and support that are both technical and pedagogical to teaching staff. This means providing mentorship and time for teachers to develop proficiency and ease with adopting ICT in their teaching and (ii) the willingness of the teachers to engage with their own professional development, that is, to spend time developing a sufficient level of digital literacy and to further explore and practise using new tools.

EQUIPPING TEACHERS WITH DIGITAL LITERACY

The complexity embedded within both the theory and pedagogy of educating pre- and in-service teachers about new technologies and their integration into teaching involves (i) their understanding of the learners they teach and how the learners learn (Hammond and Brandsford, 2005) and (ii) possessing both pedagogical content knowledge (Shulman, 1987) and technological knowledge (Mishra & Koehler, 2006). Building on Shulman's (1987) pedagogical content knowledge (PCK) model, Mishra and Koehler (2006) proposed the Technological Pedagogical Content Knowledge (TPACK) model to include technological knowledge as an integral component of teaching with technology (see Figure 1). It posits that the relationships between technology, pedagogy and content knowledge are complex and that teachers should possess a composite knowledge of these three components in order to integrate technology effectively in their teaching. For science teachers, it will mean the education of pre-service teachers in teacher education courses to develop TPACK relevant to the teaching of science with technology. One way to achieve this is through explicit demonstration, training and modelling of how technology could be used (Sprague, Kopfman & Dorsey, 1998; Willis & Cifuentes, 2005). An implication of this is that the science teacher educators (general and specialist science methods lecturers) themselves will need to have a reasonably high degree of TPACK.

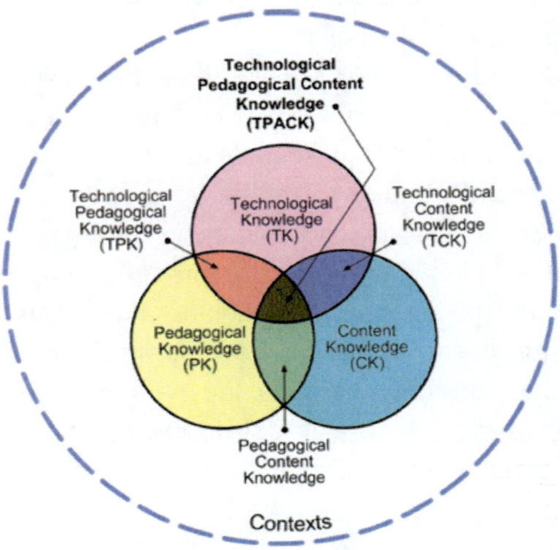

Figure 1. TPACK[5] (Mishra & Koehler 2006).

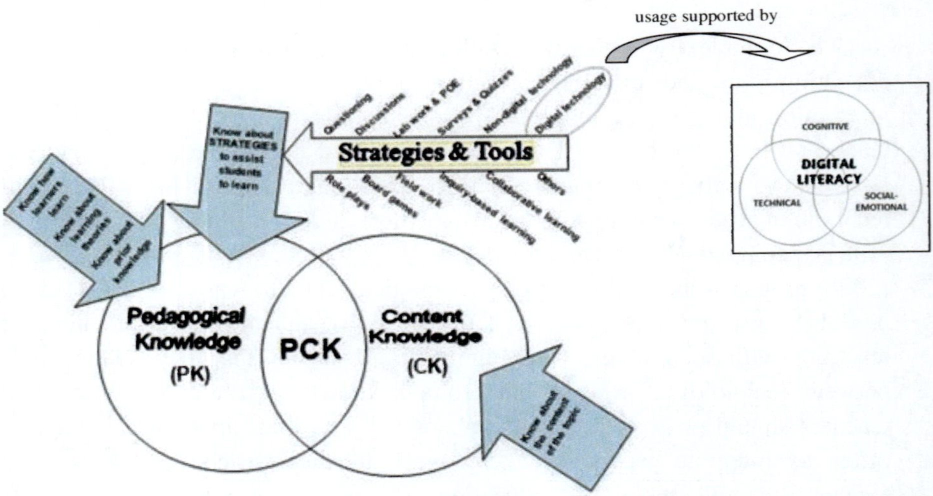

Figure 2. Framework for teachers integrating ICT into their teaching that is supported by digital literacy.

[5] Source: http://tpack.org/tpck/index.php?title=Main_Page

The technological perspective of TPACK that teachers need to develop would be the digital literacy that I have discussed in the preceding chapters. Technology is a tool to support pedagogy, just like the whiteboard, pen and paper are tools that aid with teaching and tools that allow students to represent their thinking with. The use of technology is also a strategy that assists with pedagogy. This is shown in Figure 2, which shows the place of digital technology in Shulman's (1987) pedagogical content knowledge model. As a tool, it would be sufficient for the teacher to develop the technical skills to use it as a tool technically to meet a specific purpose. (S)he does not necessarily have to make connections between the cognitive and social-emotional dimensions of digital literacy or the multiple literacies within them. Digital literacy, however, is a long-term investment. Its development is continuous and additive, building on knowledge and skills gained from previous interactions with digital technologies. It equips the science teacher with a holistic view of the forms and functions of digital technologies and the ability to embrace 'mutliliteracies' as well as 'multiple literacies' approaches to the teaching and learning of science. As discussed in Chapter 4, multiliteracies in science learning is associated with the multimodal ways of representations, these being linguistic, audio, visual, spatial, gestural and multimodal modes. Multiple literacies include a range of literacies that support ICT use, such as critical literacy, branching literacy, information literacy, social-emotional literacy etc (see Chapter 3, Figure 7). Adopting a multiliteracies framework caters for the diversity of students with preferences for different modes of learning. Adwell et al. (2007) stated that multiliteracies acknowledge the variations in students' strengths and weaknesses, similar to Gardner's (1983, 1999) multiple intelligences theory. In the multiple intelligences theory, Howard Gardner proposed nine intelligences that people posses, often in combination. These are:

- Interpersonal – reacting with others with sensitivity
- Linguistic – ability to use words and language
- Logical/Mathematical – mathematical and scientific reasoning
- Naturalistic – curious about the natural world
- Intrapersonal – knowledge of own thinking and self-reflective capacities
- Spatial – visual ability to create mental images and model
- Musical – sensitivity to rhythm, beats and patterns in tones

- Kinesthetic – physical movements and use of the physical body to learn
- Existential – asking big picture questions

Multiliteracies and multiple literacies enabled by ICT provide teachers and students with various means of understanding science concepts and creating new meanings from what has been learned, catering for the different intelligences advocated by Howard Gardner. For example, gestural/embodied modes of representations would cater for students with kinesthetic intelligence while audio modes of representations would cater for interpersonal/musical/linguistic intelligences and visual/spatial modes of representations would cater for mathematical/spatial intelligences. In practice it is likely that most students would possess a combination of intelligences, hence teachers should make use of single and multi-modes of representations as well as multiple representations to teach science. Digitally literate science teachers with an understanding of multimodality and multiple literacies across the three dimensions (cognitive, technical and social-emotional) of digital literacy are better equipped to design pedagogically sound learning materials for their students.

Science Teachers' ICT Professional Learning

The use of an appropriate range of tools and strategies, including using relevant and authentic examples to engage the students in conceptualising science knowledge, is part of the pedagogical knowledge that the science teacher needs to engage with. The more digitally literate a teacher is, the more confident (s)he will be with using digital technologies in his/her teaching. As digital literacy skills and knowledge are transferable across platforms and applications, developing digitally literate teachers would prepare them to integrate ICT under most circumstances, for example in schools with few computers, the use of mobile technologies or adopting new technologies.

Providing ICT professional education opportunities and time to practise and familiarise themselves with a range of technologies useful for science teaching and learning is imperative to developing teachers' digital literacy. As it is often the case that several software packages could do the same task, teachers need to be able to critically select the most appropriate one for use in order to achieve targeted learning goals. For example, choosing between *Prezi* or *PowerPoint* as a presentation tool to demonstrate a science concept would depend on the capabilities of each tool in achieving the purpose and how well the teacher understands the differences between them. The zooming effect of *Prezi* could be made use of to highlight specific aspects of a concept or its frame effect could be useful for showing layers of the concept, for example, organs, tissues and cells. Another example of selecting the appropriate tool to achieve a learning goal is getting students to create an ePortfolio of science artefacts that shows their development of knowledge for a science topic. These artefacts could include a concept map, an experimental report, a video creation and a research assignment. It is possible to choose from a variety of open sourced and commercial ePortfolio platforms, for example *Mahara*, *GoogleApps for ePortfolio* and *Sakai CLE*. Teachers confident with their own digital skills and knowledge would be more open and allow students to choose from a variety of similar software packages to create their ePortfolio, for example they could use *Mahara* or *GoogleApps*, both of which are open-sourced and cost-free. In using online resources such as these, educating students to work safely online is an essential part of the science teachers' role in embracing ICT in their teaching.

A willingness to invest time and persist with the development of ICT skills and knowledge, guided by mentorship from colleagues with expertise in technology use and supported by the school leaders in providing the necessary technical assistance so that problems are fixed immediately, would assist greatly with the development of digital literacy capacities in teachers. When teachers feel confident and comfortable with using technology, they will be able to assist with their students' technical problems without too much effort and hence will not be distracted from the teaching of the content that students need to learn. In this way, they are able to increase the germane load of their students' cognition through the preparation of engaging and relevant activities.

TEACHING SCIENCE

For science teachers, an understanding of their students' prior knowledge (preconceptions) and their commonly held misconceptions would be useful for the design of effective ICT-integrated lessons. Students' prior knowledge of phenomena is an important part of how they come to understand school science (Driver 1989; Driver, Squires, Rushworth & Wood-Robinson, 1994).). They construct meanings of phenomena in ways that suit their own experiences and expectations. Their interpretations of how the world works often do not accord with accepted scientific views and teachers who teach 'objectively' do not always recognise the cognitive conflict that has to be resolved within a student's mind – the conflict between the science taught at school and the science that has already been constructed out of their own real life experiences. Preconceptions that are non-concordant with scientific explanations or theories are classified by scientists as misconceptions. Misconceptions are often also referred to as alternative conceptions, that is, the reasoning about a science concept that differs from the acceptable body of scientific knowledge. As the term 'alternative conceptions' implies, misconceptions reflect processes of logical reasoning which in the student's mind make sense, just not the sense of accepted scientific logic, because the evidence on which they are based is not that of standard science. Duit (2002) has compiled a bibliography based on a collection of some 6000 papers that are largely related to students' pre-instructional (alternative) conceptions. Misconceptions in almost every area of school science have been identified from the extensive research that has been carried out in this area over the last four decades.

Probing with Digital Technologies

Figure 3 shows the results of a hand-written probe that was administered to years 8, 9 and 10 students. The probe (see Appendix 1) investigated students' prior knowledge about sugar dissolving in water and how they viewed the phenomenon at the microscopic level. They were also asked to describe what happens at the macroscopic (observable) level when a spoon of sugar was stirred into a beaker/bowl of water. The figure shows a variety of visual representations depicting the arrangement of sugar particles relative to the water particles. Almost all of the written explanations for 'describe what you see when sugar is stirred into water' were just 'dissolved' or variations of the same meaning. In this respect, the depth required to get a better insight into the students' thinking is missing. To get a deeper insight into what the students are thinking when they draw these diagrams in future, the students could voice record their explanations as they draw the sugar dissolving in water, just like in the example of Charles, the teacher who used *OneNote* to get his students to explain while drawing the structure of a cell (see Chapter 6, Example 3). A digitally captured record of what the students are thinking prior to the teaching of the chemistry behind sugar dissolving in water means that the students could re-do this activity at the end of the unit and compare their responses with what they had thought at the beginning of the unit. Alternatively, students could make use of animation software to demonstrate how the sugar has dissolved. Apart from the teacher's explanation, another source of information that students could use to learn about the dissolution of sugar is from the American Chemistry Society website at http://www.middleschoolchemistry.com/lessonplans/ where multimodality is a feature of the science resources at this site. Other means of probing with ICT is through interactive quizzes or online discussions.

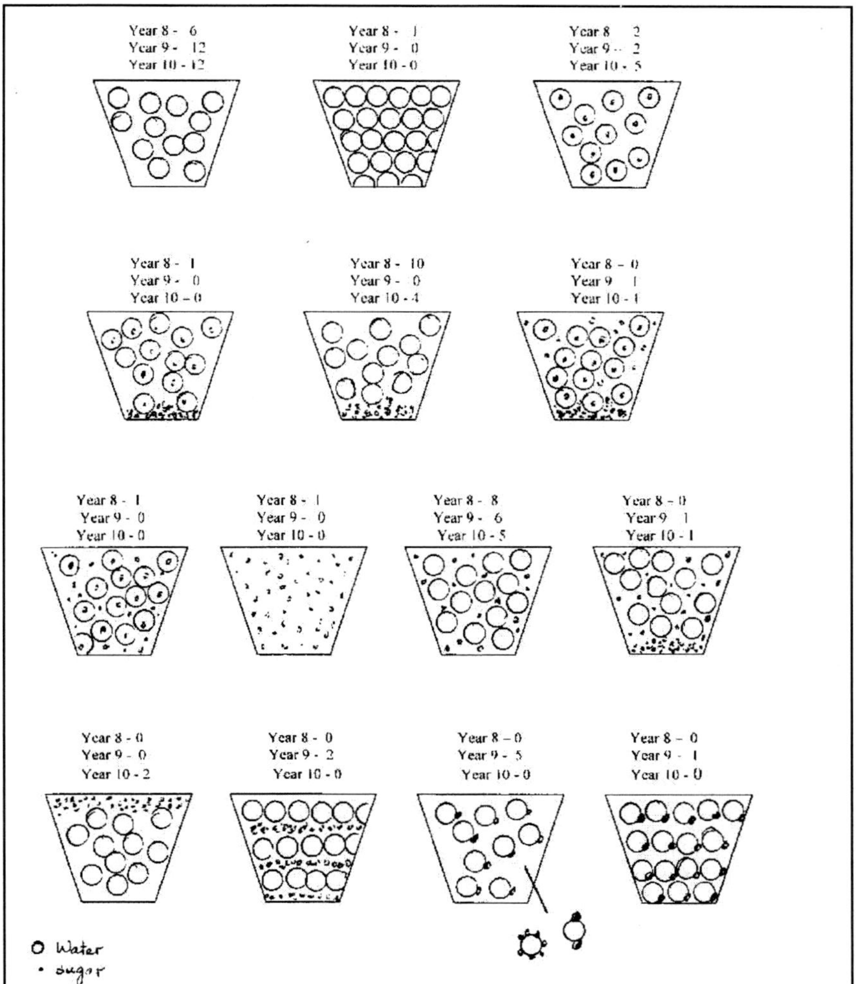

Figure 3. Students' preconceptions of what happens to sugar stirred into water.

ICT INTEGRATION MODEL FOR SCIENCE TEACHERS

For instructors (teacher educators, pre- and in-service teachers) planning to integrate ICT into their teaching, Roblyer and Doering (2010, p. 50-51) has proposed a Technology Integration Planning (TIP) model that could be adopted. There are six phases embedded in this model (i) assess technological pedagogical content knowledge (ii) determine relative advantage (iii) decide on objectives and assessment (iv) design integration strategies (v) prepare the instructional environment and (vi) evaluate and revise integration strategies. I will adapt this model in describing the digital literacy dimension of integrating ICT into the planning of teaching, using as an example the topic of *nanotechnology* for middle to upper secondary school students' learning. In doing so, the six phases will be re-ordered as:

1. Assess the teacher's own content knowledge, strategies to engage students with and his/her digital literacy proficiency in using identified technologies
2. Decide on objectives and assessment
3. Design integration strategies
4. Determine relative advantage
5. Prepare the instructional environment, including planning the time and small tasks to develop students' digital literacy proficiency in using identified technologies to learn the topic
6. Evaluate and revise integration strategies

Assess the Teacher's Own Content Knowledge, Strategies to Engage Students with and His/Her Digital Literacy Proficiency in Using Identified Technologies

The teacher assesses his/her content knowledge of the topic on nanotechnology and explores strategies, including the use of ICT, to implement in the teaching. As part of the exercise, the teacher will need to research for information from different sources such as text books and Internet sites. (S)he will keep a repository of useful sites and other resources such as animations, videos and images that could be used for the teaching. By identifying keywords and mapping them into a concept map using *Inspiration* or similar software, the teacher would be able to have an overview of what (s)he knows, where gaps are and how teaching the topic would look like.

Figure 4 shows an example of a framework for teaching and learning about nano-science and technology that is suitable for secondary students or tertiary students beginning a course on nanotechnology. The map draws together some of the basic concepts of nanotechnology and shows their relationships. The main concepts include: (i) the scale of 'nano' size and (ii) size and structure could alter properties, hence rearranging the same type of atoms into different structures could produce different products with different properties (for example, see Table 1 for the different forms of carbon only materials and their uses). In the concept map shown in Figure 4, the circles represent possible ICT integration for teaching the topic on nanotechnology.

Table 1. Carbon allotropes (Ng, 2009, p.19)

Carbon-only materials	Diamond	Graphite	Bucky ball	Carbon nanotube
	Natural		Synthetic (nanotechnology)	
Structure	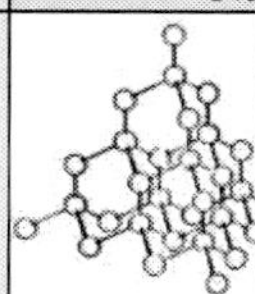	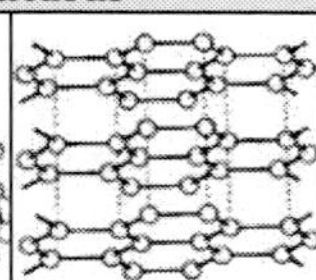	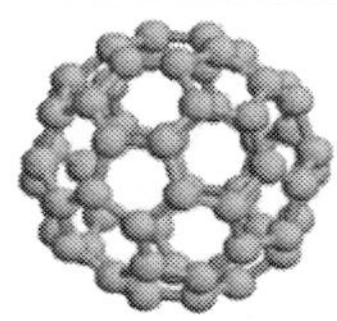	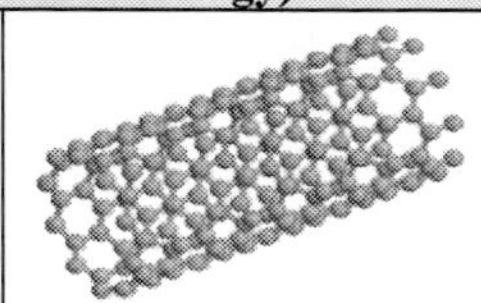
Examples of uses	- Jewellery - Abrasive e.g. for grinding and cutting hard objects	- Pencil lead - Lubricant - Electrodes in cells/batteries	- Traps drugs or fuel for slow release in systems - Super molecular ball-bearings as lubricants - As catalysts and superconductors	- Very strong fibre - Electronics (good conductors of heat and electricity) - As sensors injected into cells for cancer drugs

In this phase, the teacher reflects on his/her own prior knowledge, identify gaps, research for more information, identify strategies to teach the main concepts, identify software that are suitable for the teaching and where the need to build on his/her digital literacy capacity lies, particularly the technical aspects of new software. Table 2 shows the components of digital literacy that are involved in each phase of the planning model.

Decide on Learning Objectives and Assessment

The objectives for teaching a topic for most teachers are usually in accordance with state/district or national curriculum framework requirements. An example is the *Victorian Essential Learning Standards* (VELS) framework where at Level 5 (years 9 and 10), the focus of science learning is on *Science knowledge and understanding* and *Science at work*. The outline in Appendix II shows an example of the objectives, activities and suggested assessments that are suited for VELS, particularly adopting the interdisciplinary approach to learning science. In the case of nanotechnology, the discipline areas of biological, chemical and physical sciences are integrated.

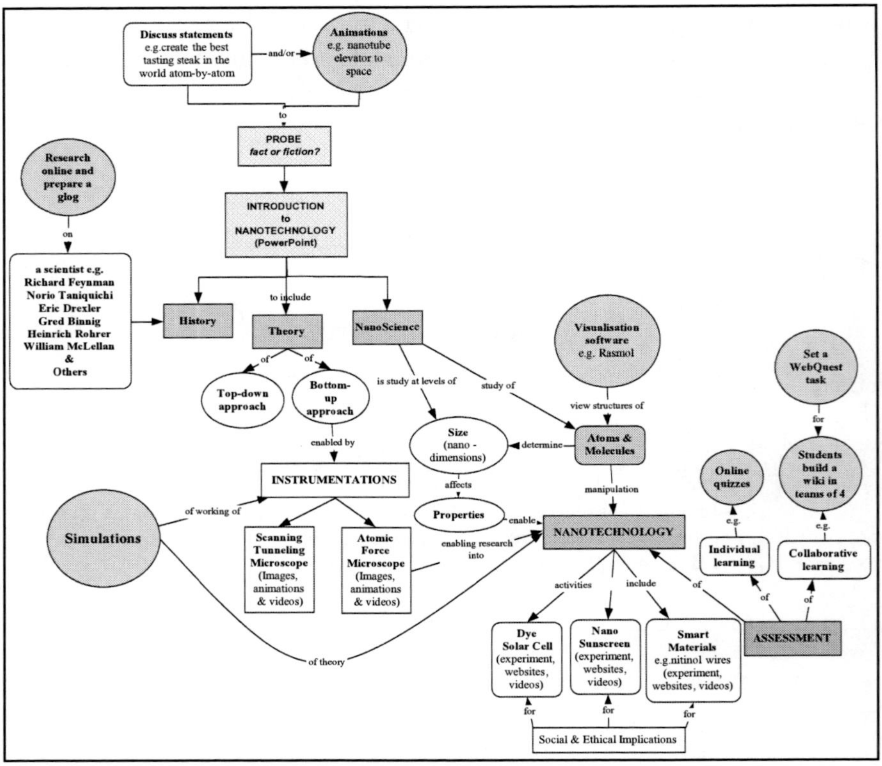

Figure 4. Concept map of the framework for learning about nano-science and technology (modified from Ng, 2009).

As shown in Appendix II, the objectives include learning about specific content related to nanotechnology supported by different pedagogical strategies, one of which is the use of ICT in the learning.

Table 2. Planning the integration of ICT into a nanotechnology learning module and the components of digital literacy involved in each phase of the planning model

Stage of planning	Activities undertaken	Digital literacy components involved
Assess the teacher's own content knowledge, practices to engage students and digital literacy proficiency in using identified technologies	Research for information and useful resources e.g. videos, images and animations	Information literacy Branching literacy Critical literacy Multiliteracies (multimodality perspective)
	Identify keywords for content, activities and ICT-enabled learning; build a visual plan e.g. concept map	Operational literacy Reproduction literacy Critical literacy
Decide on objectives and assessment	Relate to the state or national curriculum frameworks to decide on objectives and assessments.	Information literacy Critical literacy (in reading online materials)
Design integration strategies	Integrate relevant ICT tools and resources in helping students to learn and in assessing them.	Information literacy Critical literacy Multiliteracies (multimodality perspective)
Determine relative advantage	Assess the types and level of difficulty of ICT resources ensuring that they do not contain many extraneous elements to overload the working memory while learning; catering for various abilities	Operational literacy Critical literacy
Prepare the instructional environment, including planning the time and small tasks to develop students' digital literacy proficiency in using identified technologies to learn the topic	Prepare instructional materials and identify the types of environments; identify types of or areas in ICT that students need to use and be taught how to use them e.g. the students technical, cognitive and social-emotional literacy in creating a wiki online such as online safety and being sensitive to peer's editing.	Operational literacy (e.g. features in Wikispaces to construct a wiki; creating quizzes in SurveyMonkey, Hot Potatoes or Quia) Multiliteracies (multimodality) Critical literacy Information literacy Reproduction literacy
Evaluate and revise integration strategies	Integrate an action research and gather both quantitative and qualitative data to evaluate the teaching. Understand difficulties encountered with the various ICT use and revise appropriately	Information literacy (in generating questionnaires) Critical literacy (in evaluating data objectively) Operational literacy (in using SPSS and NVivo software packages)

Design Integration Strategies

Science curriculum is moving towards authentic and relevant learning in order to ensure that learning is meaningful and motivating for the students. Where possible, an integrated approach to the learning of science is encouraged. The topic of nanotechnology lends itself well as an integrated learning topic where the VELS domains are satisfied as follows:

- *Interpersonal Development* where students build social relationships by working in teams to design and/or conduct experiments or carry out online discussions in small groups or with the whole class on issues related to the development of nanotechnology in society
- *Personal Learning* where students learn individually and manage his/her own personal learning. They are responsible for carrying out their designated roles in team-based activities, for example, the team building a wiki based on a WebQuest task related to microscopy or to the applications of nanotechnology. They are individually responsible for submitting required assignments on time and participating actively in class discussions and other activities. They develop their communication skills online or in the physical classroom. They communicate verbally (discussions, oral presentations), in written form (reports and essays), graphically (concept maps, animations) and mathematically (nanoscale, data collection, graphs in practical reports). They listen and respond to small teams and whole class discussions. To further develop their communicative skills, they could organise a nanotechnology 'fair' to teach younger students, either face-to-face or online, about nanotechnology.
- *Thinking Processes* where they make use of and further develop their reasoning, processing and inquiry skills, their creativity and ability to reflect and evaluate through metacognition. Specific electronic 'thinking tools' created by the teacher that embraces Bloom's taxonomy and/or De Bono's six Thinking Hats would guide students through their thinking processes and further develop their metacognition abilities. In terms of nanotechnology, students are required to go outside of their normal range of experiences to develop understanding of abstract concepts and be able to distinguish between facts and fiction related to nano-science and technology. They are able to reflect on and evaluate information to assess the impact of nanotechnology on society. They apply science skills and processes in

designing controlled experiments, such as the use of ultraviolet beads to test the effectiveness of nano-sunscreen lotions, and discuss the limitations of the experimental research. They apply higher order thinking skills in collaborative project work where they will need to search for, evaluate and synthesise new materials with their peers.

- *Design and Create* where they investigate, design, create and evaluate a constructed product. For example they could research and design a model of a nanobot and justify their creation or they could research and design an animated model of a nano-factory showing atom-by-atom building of a useful material. Both are futuristic scenarios but provide opportunities for students to be creative and innovative in designing and constructing hypothetical models that are justified.
- *Use ICT* where they use digital technologies for creating, communicating and visualising thinking. For example, they use *Rasmol* as a visualising tool for nanotechnology-related structures and develop skills in obtaining pdb files for a range of structures from the Web. They use these *Rasmol*-generated structures as images to demonstrate their thinking in reports. They use thinking/concept mapping tools to demonstrate understanding and use *PowerPoint/Prezi* and other multimedia software to communicate understandings. They engage in a WebQuest group task set by the teacher on the the use of nanotechnology in medicine, the food industry, the cosmetic and chemical industry, focusing on the advantages, risks and ethical issues in each area. Each member in the team of 4 could be responsible for one of these areas to research on and find information. They create a wiki with their group members to diplay the group's solution to the WebQuest. There are also other proposed uses of ICT in teaching this topic as shown in Figure 4.
- *Civics and Citizenship* where the students explore the responsibilities of global citizenship (individuals, organisations and governments) and the roles and responsibilities of companies, producers and consumers in relation to sustainability. They explore ways in which countries work together to protect the environment. As discussed in Chapter 2, developing scientific literacy is to enable students to become informed citizens who can participate in debates relating to societal issues (moral and ethical) underpinned by science and technology. The development of nanotechnology and its applications has initiated a range of societal issues that students could debate on in class or in online discussion forums, for example the implications of extremely

> small materials in our body systems and in our environment. They read web-based materials and journal/book articles to develop understanding of the contribution to knowledge on nanotechnology by scientists, journalists, environmentalists and others from around the globe. In doing so they obtain a balanced view of the arguments and reflect on the position that they wish to adopt. These activities will develop the students' critical thinking, critical literacy and other digital literacy skills when assessing the information written or images produced on websites to argue their case.

The outline in Appendix II also shows the integration of an extensive range of ICT resources into the teaching and learning of nanotechnology. These include websites for obtaining information, *YouTube* videos for demonstrations or to reinforce concepts studied, interactive quizzes to test understanding of concepts developed, presentation tools to present research conducted on a scientist who has contributed to the work on nanotechnology, word processing tools to write reports, concept mapping tools to summarise concepts learned and to show their interrelationships and wikis to conduct collaborative work on a WebQuest. In embracing these technologies, the students will enhance all the three dimensions of digital literacy and the multiple literacies within them. The multimodality enabled by the range of ICT resources used would further assist students of different abilities and with preferences to different modes of presentations to learn the topic better.

Determine Relative Advantage

For the ICT choices made in the design and planning processes, the teacher needs to identify the benefits for using the technologies, for example whether the selected artefacts (such as images and animations) or the multimodality afforded by the selected software would build on the teacher's own verbal and written modes of information transfer to the students, and/or whether the technologies will cater for different abilities and interests. The teacher will need to assess the user-friendliness of the technology or digital resources which will influence the level of difficulty of usage. For example, to make effective use of the *Rasmol* software for molecular structure visualisation, it is necessary to have some knowledge of simple programming language as input of html commands is needed for some aspects of the visualisation, for example angles between bonds. The teacher needs to determine if this would overload the students' working memory while using the software to learn about chemical structures. This could be alleviated by allowing students time to explore the software and write simple codes through undertaking a set of simple exercises. Teaching students this aspect of the technology would also differentiate the curriclum as higher ability students who would learn the codes quite easily could make use of them to find out more about the structures that they are viewing. For resources such as websites or multimedia materials that are extremely 'busy', such as those with many links or crowded with text and images, it could become overwhelming for the students and overload their working memories with extraneous loads. The teacher has to determine if these materials are helpful for the students' learning and decide on the relative advantage that they would bring to the students' learning.

Prepare the Instructional Environment, Including Planning the Time and Small Tasks to Develop Students' Digital Literacy Proficiency in Using Identified Technologies to Learn the Topic

In this phase, the teacher prepares the instructional materials for teaching the topic and identify the types of environments for the different activities. Examples of the types of environments are: a research task on a scientist involved in nanotechnology or an online quiz will require the booking of a computer room if the students do not already have their own personal computers, experimental work environment where activities are conducted in a laboratory or outdoors (e.g. testing nano sunscreen using UV sensitive beads outdoors) or collaborative learning environments that are either physical or virtual. In planning the instructional materials, the teacher plans both for him/herself and the students the time and strategies to develop the necessary skills and knowledge in order to integrate the identified technolgies effectively. For example, if this is the first time the students are constructing a wiki, it is necessary to plan for some time to teach them how to use the application, on *Wikispaces* or *GoogleSites*, to build one. In this case, focusing on the students technical, cognitive and social-emotional literacy in creating a wiki online, for example being sensitive to peer's editing, are necessary aspects of the teaching.

Evaluate and Revise Integration Strategies

The planning of the instructional materials will identify areas that could potentially be problematic, depending on the abilities of the students in the class and the accessibility to the required hardware and/or software. In this respect, an evaluation of the planned instruction and the revision of integration strategies are necessary.

An action research approach to teaching the topic that is integrated into this last stage of the planning would help with the evaluation and revision for the next cycle of teaching. There are different versions to the definition of 'action research'. One of the definitions is that of Carr and Kemmis' (1986:162) who stated that action research is:

> …… a form of self-reflective enquiry undertaken by participants in social situations in order to improve the rationality and justice of their own practices, their understanding of these practices, and the situations in which the practices are carried out

While there are some variations in the literature of the definitions for action research and the sequences of actions, most would have the following elements in their sequences: (i) identification of a general idea (ii) an action plan (iii) data collection (iv) analysis of data and (v) further action plan. Action research is cyclical in nature and as many cycles as the teacher wishes to conduct could be undertaken, with each cycle trying out one to several new or modified activities.

Being able to research systematically one's own teaching practice and the impact it has on the students would enable the teacher to reflect objectively on the effectiveness of his/her own teaching. Data gathering for action research could include administrating questionnaires that contains both quantitative and qualitative questions to students. The former type of questions would require quantitative software tools like *Statistical Package for the Social Sciences* (SPSS) while for the latter type of questions, qualitative software packages like *NVivo* could be used. Being able to use these for data analysis would further enhance the digital literacy of the teacher. Action research data would inform how well the integration of ICT into the teaching and learning of nanotechnology has been achieved and what revisions need to take place.

CONCLUSION

The chapter has shown that digital technologies offer a wealth of opportunities and possibilities for teachers to use to support their teaching of science. The use of the technologies for teaching and learning requires varying levels of digital literacy, depending on the type of and user-friendliness of the technology used. Hence for teachers integrating ICT into their teaching, the more digitally literate the teacher is, the easier it is for him/her to do so through the transferral of skills and knowledge from one application to another, although new skills and knowledge may need to be developed with newer, less familiar technologies. Developing digital literacy to use ICT effectively in teaching requires time and effort to learn and practise. Teachers need to be supported by their institutional leaders in order to achieve this. In planning to integrate ICT into teaching, it is necessary for teachers to be able to use the tools that the students are expected to use. It is also the responsibility of the teacher to teach their students how to use new tools. While it is possible to learn from the students themselves as a two-way learning and teaching process, integration of ICT into the students' learning process will be much smoother if the teacher is skilled in adopting and implementing the technology himself/herself. Teachers should invest dedicate time in teaching the students new technologies and allowing time for them to explore them.

Teachers should also remember that ineffective use of ICT could hinder learning. By ineffective I mean poorly designed learning resources, complex software packages and/or the individual's (teacher or student) inability to make effective use of the features of the software due to unfamiliarity. A similar argument could be applied for the overuse of ICT, especially frequent use of the same type of applications or undertaking the same type of activities too often. There is usually a novelty factor associated with ICT use and unless students find that the use of ICT is helping with the learning, the novelty factor would disappear quickly. Sustainable use of ICT in science teaching and learning is only achievable if both teachers and students see the benefits with its use when compared to learning without ICT.

APPENDIX I

Sugar in Water

Chemistry at the Macroscopic level (What you observe)

- A teaspoon of sugar is stirred into a beaker of water. Describe what happens.

Chemistry at the Microscopic level (What happens at the atomic or molecular level?)

- Complete the diagram of what happens to the sugar and water particles after the sugar has been stirred into the water.

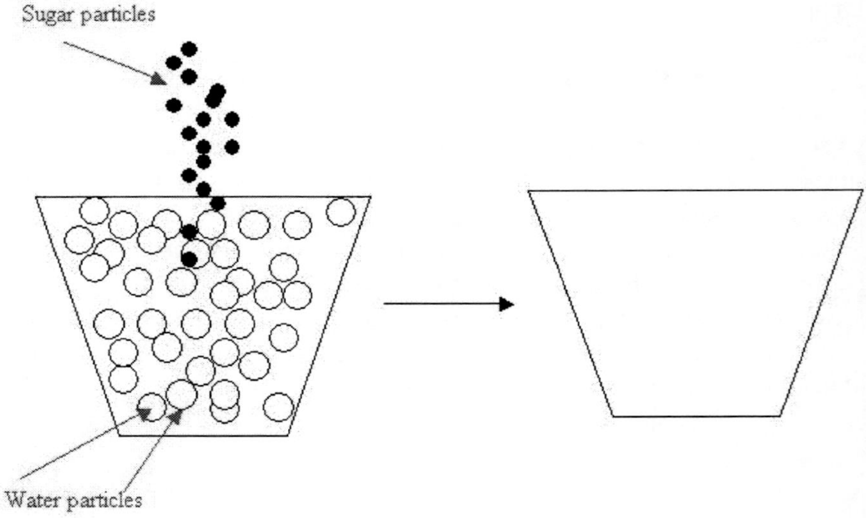

APPENDIX II

Outline of Objectives, Activities and Suggested Assessments for the Topic of Nanotechnology (Adapted from Ng, 2009)

Objectives 1: Probe for Prior Knowledge and Misconceptions and 2: Introduce Nanotechnology with an Aim to Motivate and Interest

Activities

Probe

(i) what students know: open discussions about the meaning of nanotechnology
(ii) use stimulus questions e.g. from Jones, Falvo, Taylor & Broadwell (2007). Examples of these questions are:
   - there are currently biological nanomachines that naturally exist in your body
   - NASA plans to build a space elevator that would use carbon nanotubes to move materials from Earth to outer space
   - self-cleaning toilets are now available, these toilets are made with nanotechnology that keeps the porcelain clean
   - through nanotechnology, steaks can be made atom-by-atom such that cows are no longer needed to produce the meat

Assessment

Find out where students are

Objective 3: Enable Students to Have a Sense of the Smallness of 'Nano'

Activities

From kilometre (km) to nanometer (nm). A comprehensive resource that is readily available for helping students conceptualise scale at the nanometer level can be found at:
http://www.nanosense.org/activities/sizematters/sizeandscale/SM_Lesson2Student.pdf

Define nanotechnology.

Assessment

Practical activity that sort a wide variety of materials into km, m, cm, mm and nm groups

Objective 4: Enable Students to Understand Property Changes wth Size and Structure

Activities

Properties change with structure: Show pictures of the structures of diamond, graphite, bucky ball and nanotube - these are all made of carbon atoms but are arranged differently and demonstrating different properties, hence different uses. See also http://www.science.org.au/nova/024/024print.htm

Construct a paper bucky ball with template at http://www.seed.slb.com/en/scictr/lab/buckyball/index.htm

Properties change with size of materials: Use sunscreen as example. Do the activities associated with sunscreen (see below).

Visualising molecular structures: Use RasMol (or other visualisation) software to view structures of a variety of nanotechnology related molecules, e.g. titanium oxide, zinc oxide, diamond and bucky ball.

Assessment

Online quizzes e.g. multiple choice questions, crossword puzzles and shor answer questions

Objective 5: Enable Students to Know About the Development of Instrumentations (Microscopy) in Advancing Research

Activities

Evolving microscopy from magnifying glass ---►light microscopy (compound microscope)-----►electron microscopy (scanning tunnelling microscope)---► atomic force microscope. The latter 2 enables nanoscale imaging.

Watch animation to view zooming into nanoscale levels e.g. at http://www.cneu.psu.edu/edToolsAcitivites.html

Assessment

Collaborative rsearch task to be demonstrated as a wiki e.g. a Webquest on microscopy that students work in teams of 3-4.

Objective 6: Enable Students to Have a Historical View of the Development of Nanotechnology and to See the Human Side of Nanotechnology

Activities

Explore Richard Feynman's 'marvellous biological system' idea as the origin for the concept of nanotechnology. Explore 'cells': what do they do; what do they have inside them; how the extremely tiny (and invisible) 'things' inside cells keep the whole body working

Research these people and their work and construct a timeline on the historical development of nanotechnology research, for example, Richard Feynman, William McLellan, Tom Newman, Gordon Moore, Norio Taniguchi, Tuomo Suntola, Eric Drexler, Richard Jones, Donald Huffman, Wolfgang Kraetschmer , Suomo Iijima, Richard Smalley, Gerd Binnig, Heinrich Rohrer,Calvin Quate and Christoph Gerber.

Assessment

Individual task: research and present as a PowerPoint or Prezi presentation to the class

Objective 7: Enable Students to Understand About the Applications of Nanotechnology in Their Everyday Lives

Activities

Extend imagination and thinking to artificially synthesised miniature 'things' that could work inside cells and to useful things outside the body for our day-to-day living e.g. self-cleaning glass; anti-bacterial bench top or food containers; anti-odour and stain resistant clothings, cleaner water, band-aid delivering drugs (hence no injections) and nanodiamonds (4 nm). Research and discuss some of these.

Nano-sunscreen: Investigate and compare the differences between zinc cream that stays white when applied with nano-sunscreen which disappears when applied. Learn about UVA, UVB and UVC and why they are harmful. Use UV sensitive beads to investigate which materials, e.g. paper, cloth, aluminium foil, students' sunglasses, plastic, cellophane, face foundation etc. will block out UV rays better. Include testing different brands of sunscreen. Include electromagnetic spectrum and how colours are seen.

Watch video on Sunscreen for Bottles at
http://www.csiro.au/mltimeida/pf51.html

Nitinol wires: Investigate nitinol as a 'smart material', also known as memory wire. Set shape and heat wire with Bunsen flame and when cooled reshape and test with hot water (see main text). Watch video(s) on YouTube e.g. at http://au.youtube.com/watch?v=Y7jjqXh7bB4

A useful worksheet with explanations and activities can be found at http://www.ipse.psu.edu/activities/nitinol/SmartMetal.doc

Nitinol-wired fan (or themobile): Demonstrate with hot water to turn the blades. Discuss in terms of energy transfer and energy useage.

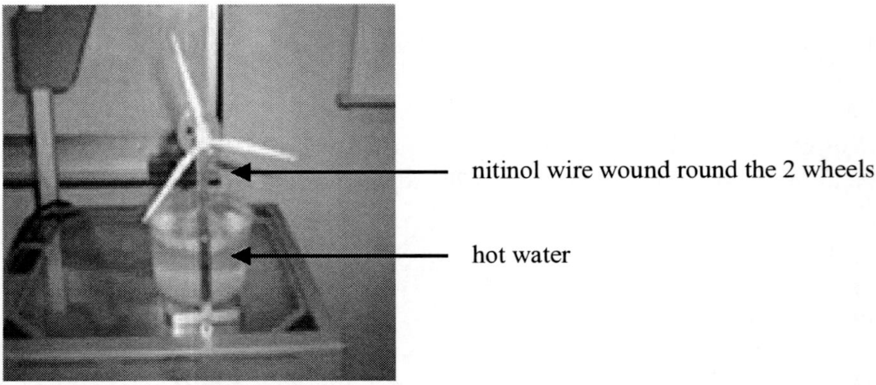

Nitinol-wired fan

As an extension, explore

(i) Dye solar cell and
(ii) Protective coating for glass surfaces. These could be purchased online at companies e.g. http://nanokote.com.au/cms/images/stories/nano_6227_glass+ceramic.pdf or http://www.enduroshield.com/en/products/enduroshield-for-glass.html

Assessment

Conduct the sunscreen and/or nitinol wire experiment and write up a report

Objective 8: Enable Students to Explore, Debate and Understand the Ethical and Social Issues Related to Nanotechnology Development

Activities

What are the implications of producing things that are really, really small? Students to research and discuss pluses and minuses/issues about nanotechnology. Examples of issues currently debated are:

- job losses if window cleaners are not needed with self-cleaning glass
- use of silver nanoparticles as anti-microbial agent
- use of nanoparticles in cosmetics and sunscreen and whether these particles are small enough to get into cells to cause harm

An informed look at these issues is necessary, hence wide reading and research is important to hear what both sides have to say.

Assessment

Debate the pros and cons of nanotech in our lives. Students then construct a concept map that shows the pros and cons of nanotech and relate the justification to content studied.

Chapter 8

Looking Forward

Introduction

This book has put forward an argument that the development of scientific literacy could be enhanced if students and teachers are digitally literate and understand that multimodality – the representation of concepts in multiple modes is inherently associated with scientific literacy. The affordances of digital technologies in enabling multimodal representations for science learning are evident in their capacities to

- contextualise learning where learning is no longer confined by the four walls of a classroom
- scaffold learning, for example through prompts, questions, hints and tutorials that include adaptive (intelligent) tutoring applications
- make visible abstract concepts and invisible entities
- offer multiple representations of a concept to cater for the range of learning preferences, interests and abilities of individuals
- enable self-directed learning and self-assessment through interactive learning objects such as simulations and quizzes
- enable collection, collation and display of primary data gathered through real-time experimentations or the manipulation of secondary (simulated) data in virtual laboratories and
- allow for continuity of learning outside of class times either through the vast amount of information and resources on the Internet or through sustained communication via learning management systems and other virtual learning communities on the Internet

But the research, as discussed in Chapter 7, seems to show that the impact of ICT on learning is still unclear and that the integration of ICT in classrooms remains variable. More recent literature review studies on the impact of ICT on science education (Hogarth, Bennett, Lubben, Campbell and Robinson, 2006; Meiers, 2009) indicated that there are potential benefits to be gained in the use of ICT, particularly in the use of simulations for science learning. These and other studies (for example Eng, 2005; Higgins, 2003; Punie, Zinnbauer & Cabrera, 2006) indicated the crucial role of the teacher in scaffolding the pedagogy and facilitating the learning at both the technical and cognitive levels when using ICT in the classroom. In addition, Hogarth et al. (2006) asserted that teachers need to be trained in the use of ICT (simulations) in order to obtain greater benefits for students' understanding. Punie et al.'s (2008) assessment of the impact of ICT on education indicated that ICT has yet to revolutionise learning and teaching. They asserted that ICT are mainly being used as tools to support existing learning processes and their associated administration but not for their transformative potential.

The effective integration of ICT into teaching and learning is complex. It is dependent on a number of things – leadership, infrastructure and set up of classrooms, the availability of computers/laptops and other digital devices, technical support, students and parents' attitudes, teaching colleagues' attitudes, the quality of the technologies available (both hardware and software) and the teachers' capabilities in adopting ICT for teaching. The complexity is further exacerbated by the fast pace of change of digital technologies, where with every change, new possibilities are opened up. There is no single or simple solution to the effective integration of ICT in teaching and learning and it takes time to develop the necessary skills to use ICT effectively (Higgins, 2003).

Educating teachers and students to be digitally literate should be the underpinning means of ensuring effective use of ICT in teaching and learning. This is similar to the need to educate students to develop (language) literacy and mathematics literacy in order to learn effectively across the school curriculum and to be functional citizens of the nation. Just like language literacy and mathematics literacy are enabling tools for learning in other subject areas, digital literacy empowers the development of understanding not only in science but in all the other subjects across the curriculum. It also empowers the individual to learn independently, effectively and safely with digital tools in informal situations while fostering the development of lifelong learning skills.

As discussed in Chapter 3, digital literacy is complex with interdependent cognitive, technical and social-emotional dimensions and the multiple literacies that are intertwined within them. Investing resources and dedicated time to develop digital literacy skills and knowledge would mean that teachers and students will not be overloaded cognitively by the technological aspects of the teaching and learning. They will be able to adopt new technologies or adapt to new versions of existing technologies without much difficulties. This is because skills and knowledge learned from various applications are transferable to other new applications. The development of digital literacy is a continuous process, it is a lifelong process that continuously adapt to changes in technologies as new ones emerge. The more digitally literate the individual, the more innovative and flexible (s)he will be with thinking about ICT and their uses to support his/her teaching, or in the case of a student, the learning.

Implications of Digital Literacy for Science Learning and Teaching

For the students who are proficient users of mobile phones and other handheld technologies, and who are frequently online in social networking communities, these abilities constitute only a portion of their digital literacy. They need to develop academically to be proficient interpreters and critical reflectors of the various forms of digital science information and resources that they interact with online and offline. In terms of educational technologies, while many students are familiar with the frequently used educational technologies such as word processors and *PowerPoint* presentation software, there are many educational tools and how to use them that they are unaware of unless being made aware of or taught by their teachers. Examples of educational technologies and how to use them include organising evidence of science learning through an ePortfolio, constructing a science wiki on wiki-enabling sites for project work, audio recording a podcast with mobile phones to explain a phenomenon underpinned by scientific theories, creating a concept map to demonstrate how well they can link key concepts together, constructing a crossword puzzle to test their peers, creating a glog online on a science concept, creating a science-related digital story and using *VoiceThread* as a presentation platform where collaborative discussions and exchange of ideas could occur. Students who are digitally literate will be able to adopt these and other new technologies easily and adapt them to their own advantage. Being digitally literate would also help the students learn science better in ICT- enabled learning environments because inhibiting factors associated with the use of the technology would not divert attention away from the content to be learned. In other words, there will not be cognitive overload and the learning would be more focused.

For the science teacher, to integrate digital technologies into his/her teaching means that (s)he will need to be sufficiently digitally literate. (S)he will need the support and time to develop both generic digital literacy skills and knowledge as well as specific ones that are distinctive to a specific software that (s)he has chosen to use. S(he) will need to plan carefully so that there is sufficient flexibility in the science curriculum and assessment to adopt ICT-based learning. Without proper preparation and implementation, ICT could impede rather than enhance learning.

CHANGING TEACHER-STUDENT POWER RELATIONSHIP

The enormous amount of freely available resources on the Internet and the affordances and opportunities enabled by digital technologies for independent learning mean that students are able to develop skills and knowledge informally at home and by connecting with peers and others in online learning communities who could help them learn. They are more autonomous learners who are able to control what they learn and when to learn them. While there are huge research gaps in the area of informal and seamless learning, it would not be unusual to find that students could be more knowledgeable in some areas than their teachers. Many students today come into the classroom with a broader set of skills and knowledge, and in the area of ICT, would be familiar with audio and visual modes of representations through their exposure to games and multimedia resources. In the classroom, more and more students are being equipped with laptops, for example the Australian *Digital Education Revolution* initiative has placed a laptop into the hands of every year 9-12 public school student, and many private schools have their own laptop programs. Many of these students' laptops are equipped with similar software that their teachers use, for example the interactive whiteboard software. As a consequence, students are empowered to learn more independently. The students today are more interactive learners rather than passive consumers of their teacher's knowledge. Hence, the dynamics in the power-relationship between teacher - student are potentially changing where the teacher is no longer the sole holder of knowledge and where learning is becoming a more democratic process.

As the trend in learning with ICT is moving towards the 'one-to-one computing' concept (see section below for more detail), information is readily accessible to students anytime and anywhere where is wireless connection. An implication is that the authority of the teacher in controlling the nature of the learning is being challenged. For exàmple, students could challenge the teacher's knowledge if other sources of information that are readily accessible in the classroom say otherwise. Another implication is that learning could be self-paced, facilitated by ICT, and teachers would need to know how to facilitate this.

DEVELOPING TRENDS IN EDUCATION

Advancement of Technologies

Adwell et al.(2007) noted the two winds of change that will impact on technologies and education of the future as (i) the rapid advancement of technologies themselves and (ii) the changing of physical classrooms to 'learning spaces' where learning spaces could be physical, virtual or a blend of both. These two changes are inter-related changes with the changing nature of learning being dependent on the advancement of technologies. For example, in a report by the Economist Intelligence Unit (2008, p. 16) on the future of higher education and how technology will shape learning, it stated that "advanced technologies will put education within the reach of many more individuals around the world, and will allow greater specialisation in curriculum and teaching methodologies than ever before." It further added that with these benefits are challenges that include having adequately supported university infrastructure and operations for the adoption of the required technologies. The transformational benefits of advanced technologies the report purports are largely attributed to the ubiquity of connectivity, that is, Internet and wireless connections would be everywhere, not just in our computers or mobile devices but woven into all fabrics of everyday living. With more advanced technology and ubiquitous connectivity, distance education would be better catered for with learning from more sophisticated learning-management systems that also provide the opportunity to collaborate with research partners from around the world.

In science research, the concept of Science 2.0 that capitalises on the capabilities of Web 2.0 has been proposed, and put in place in some cases, to redefine collaboration and the practices of scientists. Science 2.0 is characterized by open collaboration, open data and open access to publications. It refers to a new type of practice by scientists where they post raw experimental results, emerging theories, discoveries and draft papers on the Web for others to see and comment on. An example is the *OpenWetWare*[1], started by biological engineers at the Massachusetts Institute of Technology, and which aims to share information among researchers working in the areas of biology and biological engineering. As Christopher Surridge, cited in Waldrop (2008) said:

> Science happens not just because of people doing experiments but because they're discussing those experiments. Critiquing, suggesting, sharing ideas and data—this communication is the heart of science, the most powerful tool ever invented for correcting errors, building on colleagues' work and fashioning new knowledge. Although the classic peer-reviewed paper is important they're effectively just snapshots of what the authors have done and thought at this moment in time. They are not collaborative beyond that, except for rudimentary mechanisms such as citations and letters to the editor.

Advocates of Science 2.0 state that the open access nature of Science 2.0 would make scientific progress more collaborative and productive while critics are concerned with the risks of having other scientists plagiarise and/or exploit the work for their own credit.

Another level of collaboration between scientists and researchers that is enabled by Web 2.0 is through *ResearchGate*, the *Facebook* equivalent of social networking for professionals across the different discipline areas.

At the school level, online collaborative learning through social-networking, or learning-management tools is also becoming more evident. Examples include the use of *Edmodo* and *Moodle*, which are open sourced and privacy-protected systems. These systems allow students and teachers to connect at all times to share (science) content and ideas and for students to access instructions, homework, grades and school news. Mobile versions of these learning management systems are also available so that learning becomes increasingly seamless. Through these systems and other similar

[1] http://openwetware.org/wiki/Main_Page

applications on the Internet, learning science the Science 2.0 way is equally possible for school and undergraduate students in their smaller, school-based or course-based communities collaborating with similar cohorts of students around the world on science projects that seek innovative and creative solutions to posed problems.

The evolving of Web 2.0 to Web 3.0 (and Web 4.0) is already here. Web 3.0, also referred to as the 'semantic web' is able to assist learning by ensuring that the individual spends less time searching for information. By typing in one or two sentences into a Web 3.0 browser, it will analyse the content of the sentences and do an intelligent search on the Internet for all the possible responses and then organise the results for the individual. This will enable the individual to have more time absorbing and thinking about the results obtained rather than spending time reading through the large number of sites with information on the topic. How this will impact on education is yet to be seen.

Cloud Computing

The term 'cloud' is a metaphor for the Internet. Cloud computing allows the individual to use file and applications over the Internet. This means that personal files can be stored on the Internet and accessed from any computer that has access to the Internet. It means that the individual can use applications without having to download and install applications. An example is *gmail* where the software and server is managed by the cloud service provider, in this case, *Google*. Many of the applications described in the book is part of cloud computing. Examples of these include *Dropbox* or *SkyDriver*, *Prezi*, *Wikispaces* and *VoiceThread*. Science 2.0, described in the previous paragraph is about cloud computing where data is stored and shared over the Internet.

Cloud computing enables ubiquitous, on-demand access to a shared pool of resources that is Internet-based. As computing becomes increasingly personal and one-to-one (see section on 'One-to-one computing and ubiquitous learning' below), cloud computing is a trend that will become more significant in education. Students will be able to access their files and work on their homework or projects from their smartphones or tablets or gather data and upload them onto the 'cloud' without having to email large files to themselves.

Learning Spaces

An implication of the changes in the ways of learning as a result of technology is the idea of 'classrooms' transforming into 'learning spaces'. Learning spaces can be physical (classrooms, laboratories), virtual (supported by learning management systems, Web 2.0 technologies) or blended spaces. Educational institutions are increasingly re-designing learning spaces to make them more flexible and more multidimensional, for example student social learning spaces, physical learning space and course learning spaces. Formally scheduled learning takes place in physical or virtual learning spaces and less formal learning in social learning spaces where interactions with peers could also be both physical and virtual. Studies (Matthews, Adams & Gannaway, 2009; Matthews, Andrews & Adams, 2011) have shown that science students who used social learning spaces demonstrated enhanced engagement with the learning than non-users. These spaces fostered active learning, social interaction and created a better sense of belonging amongst the undergraduate students.

One-to-One Computing and Ubiquitous Learning

Mark Weiser first articulated the concept of Ubicomp or ubiquitous computing. He described three waves of computing (Weiser, 1991; Weiser and Brown, 1996). The first wave saw the use of one mainframe computer by many people, the second wave saw a one-to-one computer to human ratio where individuals each owned a desktop or laptop. We are now in the third wave of computing where one individual is serviced by many computing devices, for example a mobile phone, digital camera, GPS and game console. Schools and researchers are currently focusing on the one-to-one concept, where one student has access to an individual, often personal, computing device (such as tablet, laptop, handheld) with access to Internet information and resources. Chan, Roschelle, Hsi, Kinshuk, Sharples, Brown, Patton, Cherniavsky, Pea, Norris, Soloway, Balacheff, Scardamalia, Dillenbourg, Looi, Milrad and Hoope (2006, p. 25-26), who formed a G1:1 network argued that:

> By enabling learners to learn whenever they are curious and seamlessly switch between different contexts (such as between formal and informal contexts and between individual and social learning) and by extending the

> social spaces in which learners interact with each other, these developments, supported by theories of social learning and knowledge-building, will influence the nature, the process and the outcomes of learning. The space of social-cultural developments enabled by one-to-one TEL will unfold before us in the next decades. In particular, the ingenious or pervasive uses of these devices in some usage contexts may be close to the tipping points in terms of effecting fundamental shifts in the ways students learn in schools and outside of schools.

One-to-one computing, particularly the ownership of mobile phones as the learning tool, has the potential to close the digital divide gap. While learners around the world do not have equal access to computers, especially those in developing countries, many have access to mobile phones. The International Telecommunications Union's[2] statistics indicate that in 2010, the mobile phone subscription in developed countries was 114.2% while in the developing countries it was 70.1%. In contrast, the statistics of Internet users in developing countries was just 21.1% (this number was 68.8% in developed countries). With access to mobile phone technologies in developing countries, researchers are investigating ways of delivering education via mobile phone technologies to the people in these countries. For example, the University of Cambridge researchers in collaboration with China Mobile are examining the potential of mobile phones to deliver healthcare in China and worldwide (University of Cambridge, 2010). Other researchers such as del Carmen Valderrama Bahamóndez and Schmidt (2011) are researching ways of overcoming the scarcity of computers in schools in developing countries. They assert that the accessibility of mobile phones could change this situation and provide the potential to build feasible educational applications that enhance the learning experience of students in these countries to bridge the digital divide.

As science is all around us, having access to information anytime should motivate students to be curious about the world around them to find out about themselves, how things work and why events occur within the context that they are in. The effectiveness of the one-to-one technology-enabled learning concept would depend on the processing power of and applications in the computing device that the student holds. For example, the iPad does not support *Flash*-based learning resources, hence students will not be able to access the range of *Flash*-based science simulations on the Web. For effective and ubiquitous learning, one-to-one technology needs to be supported by the

[2] http://www.itu.int/ITU-D/ict/statistics/at_glance/KeyTelecom2010.html

suite of technologies that are dispersed throughout the physical environment (Nicholas, 2011). This, according to Weiser's (1991, 1996) third wave of computing, is the many technologies servicing one person (one to many) concept. As these technologies recede into the background of people's lives and become 'invisible', they are increasingly being used unconsciously for task completion. For example, the student may need access to the printer to print an image from his/her laptop, use the scanner to scan in an illustration, or use a mobile device (mobile phone, flipcam, digital camera) to take a video of or snapshot of a rock or plant and its surroundings to take back to class for further analysis. Using an array of mobile devices to support learning requires understanding of the transferability of files from one platform to another and compatibility issues.

CONCLUSION

One-to-one technology enabled learning with Internet access and access to a suite of 'invisible' technologies that surrounds us foster learning that is ubiquitous and 'always on'. As the trends described above continue to evolve, the reality for education is that teachers will need to embrace more technology-based instructions to teach in flexible ways that take into account students' informal learning and what that brings to the classroom. Developing digital literacy skills and knowledge in students will ensure that students use technology proficiently for academic development and are being supported in their learning in informal situations in safe and sensible ways.

Chan et al. (2006, p. 27) pointed out that despite the positives that one-to-one and mobile computing offer for learning,

> ...we need to remain cognizant of the potential negative issues associated with one-to-one TEL (technology enabled learning) discussed in the article, namely, the penetration of pervasive computing into all of one's life-spheres with potential for creating an unbalanced lifestyle; data security, integrity and privacy issues; persistent digital divide; high environmental and ecological costs of low-cost pervasive computing; and learning supported by one-to-one TEL for unethical and socially destructive purposes.

Raising awareness of these issues and how to handle them are part of developing digital literacy in students. Developing good levels of digital literacy in students (and their teachers) is an inevitable part of education. It empowers students to learn science and other subjects more effectively. It is one of the literacies that is a life skill that should be taught well.

REFERENCES

ACER (2007). *Australian students perform well in science, but many below OECD baseline.* Retrieved July 24, 2011, from *http://www.acer.edu.au/documents/MR_071204-PISA2006.pdf*

Adwell, J., Copeland, N., Forber-Pratt, A.. Gitonga, K., Kleber, K., Koplinski, C., Lawhead, M.J., Martin, R., Mattison, R. & Wilken, K. (2007). *Multiliteracies and education.* Retrieved May 23, 2011 from *http://wik.ed.uiuc.edu/index.php/Multiliteracies_and_Education.*

Aikenhead, G.S. (2003). STS education: A rose by any other name. In R. Cross (Ed.), *A vision for science education: Responding to the work of Peter J. Fensham* (pp. 59-75). New York, USA: Routledge Press.

Ainsworth, S. (1999). The functions of multiple representations. *Computers & Education*, *33*, 131–152.

Ainsworth, S., & Iacovides, I. (2005, August). *Learning by constructing self-explanation diagrams.* Paper presented at the 11th Biennial Conference of European Association for Research on Learning and Instruction, Nicosia, Cyprus.

American Association for the Advancement of Science (AAAS), (1989). *Science for all Americans: A Project 2061 report on literacy goals in science, mathematics, and technology.* Washington, DC: AAAS

American Association for the Advancement of Science (AAAS), (1993). *Benchmarks for science literacy.* New York: Oxford University Press.

Anderson, L. W. and Krathwohl, D. R. (2001). *A taxonomy for learning, teaching, and assessing: A revision of Bloom's Taxonomy of Educational Objectives.* New York: Longman.

Angelique, L. & Lim, Y.L. (2011). A comparison of students' reflective thinking across different years in a problem-based learning environment. *Instr Sci*, *39*, 171–188.

Aviram, A. & Eshet-Alkalai, Y. (2006). Towards a Theory of Digital Literacy: Three Scenarios for the Next Steps. *European Journal of Open, Distance and E-Learning.* Retrieved May 12, 2011 from *http://www.eurodl.org/materials/contrib/2006/Aharon_Aviram.htm*

Baddeley, A. D. (1992). Working memory. *Science*, *255*, 556-559.

Bardini, C., Pierce, R. & Stacey, K. (2004). Teaching linear functions in context with graphics calculators: students' responses and the impact of the approach on their use of algebraic symbols. *International Journal of Science and Mathematics Education*, *2*(3), 353-376.

Becker, H. J. (2000, January). *Findings from the teaching, learning, and computing survey: Is Larry Cuban right?* Revision of a paper written for the School Technology Leadership Conference of the Council of Chief State School Officers, Washington, D.C. Retrieved July 15, 2011 from *http://www.crito.uci.edu/tlc/findings/ccsso.pdf.*

BECTA (2006). *Safeguarding children in a digital world: Developing a strategic approach to e-safety.* Retrieved May 12, 2011 from *http://www.mmiweb.org.uk/publications/ict/safeguarding.pdf.*

Belanger, Y. (2007) The iPod revolution: Coming to a classroom near you. *Proceedings of the 2007 Organisational Systems Research Association* conference. Retrieved March 18, 2011 from *http://www.osra.org/2007/papers/brookshire.pdf.*

Bell, F. (2011). Connectivism: Its place in theory-informed research and innovation in technology-enabled learning. *International Review of Research in Open and Distance Learning*, *12* (3), 98-118.

Bennett, S., Maton, K., Kervin, L. (2008). The digital natives'debate: a critical review of the evidence. *British Journal of Educational Technology*, *39*(5), 775-786.

Berson, I. R. & Berson, M. J. (2003). Digital literacy for effective citizenship. Social *Education, 67*(3), 164-167.

Bloom B. S. (1956). *Taxonomy of educational objectives, Handbook I: The cognitive domain*. New York: David McKay Co Inc.

Bodmer, W. (1985). *The Public Understanding of Science*. London: Royal Society.

Boitshwarelo, B. (2010). Proposing an Integrated Research Framework for Connectivism: Utilising Theoretical Synergies. *International Review of Research in Open and Distance Learning*, *12*(3), 161-178.

Braithwaite, A. (2006). *New science curriculum 'more suited to the pub'.* Retrieved May 3, 2011 from *http://www.independent.co.uk/news/education/education-news/new-science-curriculum-more-suited-to-the-pub-419567.html*

Branscomb, A. W. (1981). Knowing how to know. *Science, Technology, & Human Values*, *6*(36), 5–9.

Brown, C. & Czerniewicz, L. (2010). Debunking the 'digital native': beyond digital apartheid, towards digital democracy. *Journal of Computer Assisted Learning*, *26* (5), 357–369.

Brown, J. S., Collins, A., & Duguid, P. (1989). Situated cognition and the culture of learning. *Educational Researcher*, *18*(1), 32-42.

Bruce, B.C. (1997). Literacy technologies: What stance should we take? *Journal of Literacy Research*, *29*(2), 289–309.

Bruner, J (1960). *The process of education.* Cambridge, MA.: Harvard University Press.

Bruner, J. (1966). *Toward a theory of instruction*. Cambridge, MA: Harvard University Press.

Bybee, R. (1997). *Achieving scientific literacy: From purposes to practices.* Portsmouth, NH: Heinemann.

Bybee, R. W., & DeBoer, G. (1994). Research on the goals for science education. In D. L. Gabel (Ed.), *Handbook of research on teaching and learning of science* (pp. 357-387). New York: Macmillan

Bybee, R.W., Powell, J.C.., Ellis, J.D., Giese, J.R., Parisi, L. & Singleton, L. (1991). Integrating the History and Nature of Science and Technology in Science and Social Studies Curriculum. *Science Education*, *75*(1), 143 - 155.

Caron, A., Caronia, L. & Gagné, P.(2011). Beyond mobile learning: Identity construction and the development of social awareness. In W. Ng (Ed.), *Mobile technologies and handheld devices for ubiquitous learning: Research and pedagogy* (pp. 34-57). Hershey, PA: IGI Global Publishing .

Carr, W. and Kemmis, S. (1986) *Becoming Critical. Education, knowledge and action research*, Lewes: Falmer.

Chan, T.W., Roschelle, J., Hsi, S., Kinshuk, Sharples, M., Brown, T., Patton, C., Cherniavsky, J., Pea, R., Norris, C., Soloway, E., Balacheff, N., Scardamalia, M., Dillenbourg, P., Looi, C.K., Milrad, M. & Hoope, U. (2006) One-to-one technology-enhanced learning: an opportunity for global research collaboration. *Research and Practice in Technology Enhanced Learning, 1*(1), 3–29.

Chansanchai, A. (2006). *Millennials lead the wired life: New generation digital natives in a land of digital immigrants*. Retrieved July 24, 2011 from *http://www.msnbc.msn.com/id/14560871/*

Chng, L.P, Abas, Z.W., Goolamally, N.T., Yussoff, Y. & Kaur, H, Implementation of mobile learning at the Open University Malaysia. In W. Ng (Ed.), *Mobile technologies and handheld devices for ubiquitous learning: Research and pedagogy* (pp. 170-186). Hershey, PA: IGI Global Publishing (2011).

Clark, J. M. & Paivio, A. (1991). Dual coding theory and education. *Educational Psychology Review*, *3*(3), 149-210.

Coffey, H. (nd). *Critical literacy.* Retrieved May 12, 2011 from *http://www.learnnc.org/lp/pages/4437?style=print.*

Connor, R. (2007). *Science and Technology Literature Review.* NSW, Australia: Board of Studies.

Conroy, S. (2007). *Labor's plan for cyber-safety (fact sheet).* Retrieved March 18, 2011 from *http://machinegunkeyboard.com/ausnetcensorship/labors_plan_for_cyber_safety.pdf.*

Coombs, P. H. and Ahmed, M. (1974). *Attacking Rural Poverty: How non-formal education can help.* Baltimore: John Hopkins University.

Crispin, D. and Pymm, J. (2009). Podagogy: The iPod as a learning technology. *Active Learning In Higher Education, 10*(1), 84-96.

Crook, C., Harrison, C., Farrington-Flint, L., Tomás, C. and Underwood, J. (2010). *The impact of technology: Value-added classroom practice. Final report.* UK: BECTA.

Cuban, L. (2001). *Oversold and underused: Reforming schools through technology, 1980- 2000.* Cambridge MA: Harvard University Press.

Curriculum Corporation, (1994). *Science - A curriculum profile for Australian schools.* Victoria, Australia: Curriculum Corporation.

Daniels, J.. Takach, B. S., and Varnhagen C. (2002). Getting stuck in the world wide web: The impact of design on navigation. In P. Barker and S. Rebelsky, (eds), *Proceedings of ED-MEDIA, 2001 World Conference on Educational Multimedia, Hypermedia & Telecommunications* (pp. 371-372). Denver, USA. Norfolk, VA: Association for the Advancement of Computing in Education.

Darken, R.P., Bernatovich, D., Lawson, J., & Peterson, B. (1999). Quantitative measures of presence in virtual environments: The roles of attention and spatial comprehension. *Cyberpsychology and Behavior, 2*(4), 337-347.

DeBoer, G. (2000). Scientific literacy: Another look at its historical and contemporary meanings and its relationship to science education reform. *Journal of Research in Science Teaching, 37*(6), 582–601.

Dede, C., Salzman, M.C., Bowen Loftin, R., & Sprague, D. (1999). Multisensory Immersion as a Modeling Environment for Learning Complex Scientific Concepts in N. Roberts, W. Feurzeig & B. Hunter (Eds.), *Computer Modelling and Simulation in Science Education* (pp. 282-319). Berlin: Springer.

Dekkers, J., & De Laeter, J. R. (2001). Enrolment trends in school science education in Australia. *International Journal of Science Education, 23*(5), 487-500.

del Carmen Valderrama Bahamóndez, E. & Schmidt, A. (2011). Mobile phones, developing countries and learning, In W. Ng (Ed.), *Mobile technologies and handheld devices for ubiquitous learning: Research and pedagogy* (pp. 120-137). Hershey, PA: IGI Global Publishing.

Department for Education and Skills (DfES) (1995). *Science in the National Curriculum*. London: DfES

Dewey, J. (1938). *École et société*. Paris: PUF.

Dierking, L.D., Falk, J.H., Rennie, L., Anderson, D., & Ellenbogen, K. (2003). Policy statement of the 'Informal Science Education' ad hoc committee. *Journal of Research in Science Teaching*, *40*(2), 108–111.

Dodge, B. (2007). *WebQuest.org*. Retrieved June 5, 2011, from *http://webquest.org/index.php*

Downes, S. (2005). E-learning 2.0. *eLearn Magazine*, 10. Retrieved May 12, 2011 from *http://www.4shared.com/document/FcJQ9z1z/An_introduction_to_connective_.html.*

Driver, R. (1989). Students' conceptions and the learning of science. *International Journal of Science Education 11*(5), 481-490.

Driver, R., Squires, A., Rushworth, P., & Wood-Robinson, V. (1994). *Making sense of secondary science: Research into children's ideas*. London, UK: Routledge.

Drucker, P. F. (1994). The age of social transformation. *The Atlantic Monthly (November)*, 53-80.

Duit, R. (2002). *Bibliography : Students' and teachers' conceptions and science education*. Retrieved July 2, 2011 from *http://www.ipn.uni-kiel.de/aktuell/stcse/stcse.html.*

Dwyer, D. (1994). Apple classrooms of tomorrow: What we've learned. *Educational Leadership*, *51*(7), 4-10.

Economist Intelligence Unit (2008). *The future of higher education: How technology will shape learning*. Retrieved July 7, 2011 from http://www.nmc.org/pdf/Future-of-Higher-Ed-(NMC).pdf.

Education Services Australia (ESA) (2010). *ICT professional learning: National mapping of ICT-based professional learning*. Retrieved July 2, 2011 from *http://www.deewr.gov.au/Schooling/DigitalEducationRevolution/DigitalStrategyforTeachers/Documents/ICTMappingFinalReport.pdf.*

Educational Testing Services [ETS] (2007). *Digital transformation: a framework for ICT literacy*. Princeton: ETS.

Eng, T.S. (2005). The impact of ICT on learning: a review of research. *International Education Journal*, *6*(5), 635-650.

Ericsson, K. A., & Kintsch, W. (1995). Long-term working memory. *Psychological Review*, *102*(2), 211-245.

Eshach, H. (2007). Bridging in-school and out-of-school learning: Formal, non-formal, and informal education. *Journal of Science Education and Technology*, *16*(2), 171-190.

Eshet-Alkalai, Y. (2004). Digital Literacy: a conceptual framework for survival in the digital era. *Journal of Multimedia and Hypermedia*, *13*(1), 93-106.

E-Skills UK (2009).*Technology counts: IT and Telecoms insights 2008.* Retrieved May, 12, 2011 from *http://www.nwua.ac.uk/HLSP/docs/eskillsUK_TechnologyCounts.pdf.*

Fensham. P.J . (1985). Science for all: A reflective essay. *Journal of Curriculum Studies*, *17*(4), 415-435.

Fensham, P. J. (2004). *Engagement with science: An international issue that goes beyond knowledge.* Paper presented at the Science and Mathematics Education Conference (SMEC) Conference, Dublin. Retrieved July 24, 2011 from *www.dcu.ie/smec/plenary/Fensham,%20Peter.pdf* .

Finkelhor, D. Mitchell, K. & Wolak, J. (2000). *Online victimization: A report on the nation's youth.* National Center for Missing and Exploited Children Bulletin, Alexandria, VA. :US Department of Justice, Washington, DC.

Finn, M. & Vandenham, N. (2004). The handheld classroom: Educational implications of mobile computing. *Australian Journal of Emerging Technologies and Society*, *2*(1), 1-15.

Fischer, G. & Scharff, E. (1998). Learning technologies in support of self-directed learning. *Journal of Interactive Media in Education*, *98*(4). Retrieved May 3, 2011 from *http://jime.open.ac.uk/article/1998-4/27*

Fluck, A. E. (2008). M-Learning Landscapes: e-Learning for ubiquitous school science education. *Australian Educational Computing*, *23*(1), 24-32.

Fortuna, C., Henderson, S., McLuckie, J., Rodrigues, S., Syme-Smith, L., Taylor, N. & Williamson, G. (2010) in S. Rodrigues (Ed.), *Multiple Literacy and Science Education: ICTs in Formal and Informal Learning Environments* (pp. 1-10). Hershey, PA: IGI Global Publishing.

Freebody, P. & Luke, A. (1990). Literacies programs: Debates and demands in cultural context. *Prospect: Australian Journal of TESOL*, *5*(7), 7-16.

Freeman, J., Lessiter, J., & IJsselsteijn, W.(2001). An introduction to presence: A sense of being there in a mediated environment. *The Psychologist*, *14*(4), 190-194.

Gardner, H. (1983). *Frames of mind: The theory of multiple intelligences*. New York: Basic Books.

Gardner, H. (1999). *Intelligence reframed: Multiple intelligences for the 21st century*. New York: Basic Books.

Garrison, D. R. (1992). Critical thinking and self directed learning in adult education: An analysis of responsibility and control issues. *Adult Education Quarterly, 42*(3), 136-148.

Goodrum, D., Hackling, M. & Rennie, L. (2001). *The status and quality of teaching and learning of science in Australian schools*. Canberra: Department of Education, Training and Youth Affairs.

Grant, L., Daanen, H. & Rudd, T. (2007). *Mobile, collaborative and location-based learning: a case study of the MobiMissions prototype*. UK: FutureLab.

Green, R. E., Krause, J, Ptak, S. E., Briggs, A. W., Ronan, M. T., Simons, J. F., et al. (2006). *Analysis of one million base pairs of Neanderthal DNA. Nature. 16*, 330–336. Retrieved July 24, 2011 from *http://www.nature.com/nature/journal/v444/n7117/abs/nature05336.html.*

Hague, C. & Payton, S. (2010). *Digital literacy across the curriculum*. UK: Futurelab.

Hammond, L.D., & Brandsford, J. (2005). *Preparing teachers for a changing world*. San Francisco, CA: Jossey-Bass Wiley.

Hanewald, R. (2008). Confronting the Pedagogical Challenge of Cyber Safety. *Australian Journal of Teacher Education*, *33*(3), 1-16.

Hartnell-Young, E. & Heym, N. (2008). *How mobile phones help learning in secondary schools*. University of Nottingham, UK: Learning Sciences Research Institute.

Hazen, R. (2002). *Why Should You Be Scientifically Literate*? Retrieved April 30, 2011 from http://www.actionbioscience.org/newfrontiers/hazen.html.

Helsper, E. & Eynon, R. (2010). Digital natives: where is the evidence? *British Educational Research Journal*, *36*(3), 503-520.

Hew, K.F., & Brush, T. (2007). Integrating technology into K-12 teaching and learning: Current knowledge gaps and recommendations for future research. *Educational Technology Research and Development*, *55*(3), 223-252.

Higgins, S. (2003). *Does ICT improve learning and teaching in schools? A professional user review of UK research*. UK: British Educational Research Association.

Hodson, D. (2008). *Towards scientific literacy: A teachers' guide to the history, philosophy and sociology of science*. Rotterdam, Netherlands: Sense Publishers.

Hoffman, J.L., Wu, H.K., Krajcik, J.S. & Soloway, E. (2003). The nature of middle schools learners' science content understandings with the use of on-line resources. *Journal of Research in Science Teaching*, *40*(3), 323-346.

Hogarth, S., Bennett, J., Lubben, F., Campbell, B. & Robinson, A. (2006) ICT in science teaching. Technical Report. In: *Research Evidence in Education Library*. London: EPPI-Centre, Social Science Research Unit, Institute of Education, University of London.

Honeyman, B. (1998). Non-formal and formal learning interactions: New directions for scientificic and technological literacy. *Connect*, *23*(1), 1-2.

Horton, W. (2000). *Designing web-based training*. New York: Wiley & Sons.

Hurd, P. D. (1958). Science literacy: Its meaning for American schools. *Educational Leadership*, *16*(1), 13-16.

Ito, M., Horst, H., Bittanti, M., Boyd, D., Herr-Stephenson, B., Lange, P. G., Pascoe, C. J., Robinson, L., Baumer, S., Cody, R., Mahendran, D., Martínez, K., Perkel, D., Sims, C. & Tripp, L. (2008). *Living and Learning with New Media: Summary of Findings from the Digital Youth Project*. Retrieved July 20, 2011 from *http://digitalyouth.ischool.berkeley.edu/files/report/digitalyouth-WhitePaper.pdf.*

Jones, L. (1980). *Great expectations: America and the baby boom generation*. New York: Coward, McCann & Geoghegan.

Jones, M. G., Falvo, M.R., Taylor, A.R. & Broadwell, B.P. (2007). *Nanoscale Science: Activities for Grades 6-12*. US: NSTA Press.

Kane, S. (not dated). Baby boomers. Retrieved May 12, 2011 from *http://legalcareers.about.com/od/practicetips/a/Babyboomers.htm.*

Kaplan, A. M. & Haenlein, M. (2010). Users of the world, unite! The challenges and opportunities of Social Media. *Business Horizons*, *53*(1), 59–68.

Karsenti, T., Brodeur, M., Deaudelin, C., Larose, F. & Tardif, M. (2002). *Integrating ICTs in teacher tr@ining: a challenging balance*. Paper presented at the Pan-Canadian Education Research Agenda Symposium: Information Technology and Learning, Montreal, Quebec. Retrieved July 2, 2011 from *http://www.cesc.ca/pceradocs/2002/papers/TKarsenti_TEN.pdf.*

Kellner, D. (2004). Technological Transformation, Multiple Literacies, and the Re-visioning of Education. *E–Learning*, *1*(1), 9-37.

Kennedy, G., Judd, T.S., Churchward, A., Gray, K. & Krause, K. (2008). First year students' experiences with technology: Are they really digital natives? *Australian Journal of Educational Technology*, *24*(1), 108-122.

Keogh, K. (2011). Using mobile phones for teaching, learning and assessing Irish in Ireland: Processes, Benefits and Challenges. In W. Ng (Ed.), *Mobile technologies and handheld devices for ubiquitous learning: Research and pedagogy* (pp. 236-258). Hershey, PA: IGI Global Publishing.

Klopfer, E, Squire, K & Jenkins, H (2002). Environmental detectives: PDAs as a window into a virtual simulated world. *Proceedings of IEEE International Workshop on Wireless and Mobile Technologies in Education* (pp. 95-98). Vaxjo, Sweden: IEEE Computer Society.

Knain, E. (2006). Achieving science literacy through transformation of multimodal textual resources. *Sci Ed, 90*(4), 656–659.

Koballa, T., Kemp, A. & Evans. R. (1997). The spectrum of scientific literacy. *Science Teacher*, *64*(7), 27–31.

Kozma, R. B. (2003). Global perspectives: innovative technology integration practices from around the world. *Learning & Leading with Technology*, *31*(2), 6-9, 52-54.

Lankshear, C. & Knobel, M. (1998). *Critical literacy and new technologies*. Paper presented at the American Education Research Association annual conference, San Diego. Retrieved May 12, 2011 from *http://everydayliteracies.net/critlitnewtechs.html*

Lankshear, C. & Knobel, M. (2003). *New Literacies: Changing Knowledge and ClassroomPractice*. Buckingham, UK: Open University Press.

Laugksch, R. (2000). Scientific literacy: A conceptual overview. *Science Education*, *84*(1), 71–94.

Lave, J., & Wenger, E. (1990). *Situated Learning: Legitimate Periperal Participation*. Cambridge, UK: Cambridge University Press.

Lee, J. J. & Hsu, Y. (2002). Web navigation: The role of metaphor, concept map and individual differences. In P. Barker and S. Rebelsky, (Eds.), *Proceedings of ED-MEDIA, 2001 World Conference on Educational Multimedia, Hypermedia & Telecommunications* (pp. 1000-1001). Denver, USA. Norfolk, VA: Association for the Advancement of Computing in Education.

Leu, Jr.,D. J. (2002). The new literacies: Research on reading instruction with the Internet and other digital technologies. In J. Samuels and A. E.

Farstrup (Eds.), *What research has to say about reading instruction* (pp. 310- 336). Newark, DE: International Reading Association.

Leu, Jr., D.J., Kinzer , C.K., Coiro, J., & Cammack, D.W. (2004). Toward a Theory of New Literacies Emerging From the Internet and Other Information and Communication Technologies. In R.B. Ruddell & N.J. Unrau (Eds.), *Theoretical Models and Processes of Reading* (pp. 1570-1613). Newark, DE: International Reading Association.

Lombard, M., & Ditton, T. (1997). At the heart of it all: The concept of presence. *Journal of Computer-Mediated Communication*, *3*(2), Retrieved May 12, 2011 from http://jcmc.indiana.edu/vol3/issue2/lombard.html

Luke, A. (2000). Critical literacy in Australia: A matter of context and standpoint. *Journal of adolescent and adult literacy*, *43*(5), 448-461.

Lyons, T. (2006). The puzzle of falling enrolments in physics and chemistry courses: Putting some pieces together. *Research in Science Education, 36*(3), 285-311.

McMahon, M., & Pospisil, R. (2005). Laptops for a digital lifestyle: Millennial students and wireless mobile technologies. In H. Goss (Ed.), ASCILITE 2005 Balance, Fidelity, Mobility: maintaining the momentum? *Proceedings of the 22nd annual conference of the Australasian Society for Computers in Learning in Tertiary Education* (Vol. 2, pp. 421-431). Brisbane: ASCILITE.

Markauskaite, L. (2006). Towards an integrated analytical framework of information and communications technology literacy: From intended to implemented and achieved dimensions. *Information Research*, *11*(3), 252–280.

Martin, A. (2005). DigEuLit – a European Framework for Digital Literacy: a Progress Report. *Journal of eLiteracy*, *2*, 130-136.

Martin, A., & Grudziecki, J. (2006). *DigEuLit: Concepts and tools for digital literacy development.* Retrieved April 30, 2011 from: *http://www.ics.heacademy.ac.uk/italics/vol5iss4/martin-grudziecki.pdf*

Matthews, C. M. (2007). *Science,engineering, and mathematics education: Status and issues*. US: Congressional Research Services. Retrieved July 20, 2011 from *http://www.fas.org/sgp/crs/misc/98-871.pdf*

Matthews, K. E., Adams, P., & Gannaway, D. (2009). *The impact of social learning spaces on student engagement.* Paper presented at the 12th annual Pacific Rim First Year in Higher Education conference, Townsville, Qld, June 2009. Retrieved July 20, 2011, from *http://www.fyhe.qut.edu.au/past_papers/papers09/content/pdf/3A.pdf.*

Matthews, K. E., Andrews, V. & Adams, P. (2011). Social learning spaces and student engagement. *Higher Education Research and Development, 30*(2), 105-120.

Mayer, R. E. & Sims, V. K. (1994). For whom is a picture worth a thousand words? Extensions of a dual-coding theory of multimedia learning. *Journal of Educational Psychology*, *86*(3), 389-401.

Mayer, R. E. (1997). Multimedia learning: Are we asking the right questions? *Educational Psychologist, 32*(1), 1-19.

McKenzie, J. (1998). Grazing the Net: Raising a generation of free-range students. *Phi Delta Kappan, 79* (September), 26-31.

Meiers, M. (2009). The use of ICTs in schools in the digital age: what does the research say? *NSWIT Digest*, *2009*(1). Retrieved May 5, 2011 from *http://www.nswteachers. nsw.edu.au.*

Millar, R. (1996). Towards a science curriculum for public understanding. *School Science Review, 77*(280), 7-18.

Miller, G.A. (1956). The magical number seven plus or minus two: some limits on our capacity for processing information. *Psychological Review*, *63*(2), 81–97.

Miller, J. D. (1983). Scientific literacy: A conceptual and empirical review. *Daedalus*, *112*(2), 29–48.

Mishra, P. & Koehler, M.J., (2006). Technological pedagogical content knowledge: A framework for teacher knowledge. *Teachers College Record*, *108*(6), 1017-1054.

Mistler-Jackson, M & Songer, N.B. (2000). Student motivation and Internet technology: Are students empowered to learn science? *Journal of Research in Science Teaching*, *37*(5), 459-479.

Moreno, R., & Mayer, R. E. (2000). A learner-centered approach to multimedia explanations: Deriving instructional design principles from cognitive theory. *Interactive Multimedia Electronic Journal of Computer Enhanced Learning*. Retrieved March 22, 2011, from http://imej.wfu.edu/articles/2000/2/05/index.asp.

Mulhearn, G. & Hill, S. (2007). Children of the New Millennium: Research and Professional Learning in Practice. *Journal of Australian Research in Early Childhood Education*, *14*(1), 57–66.

Murray, C. (2011). Imagine mobile learning in your pocket. In W. Ng (Ed), *Mobile technologies and handheld devices for ubiquitous learning: Research and pedagogy*, pp. 209-236. Hershey, PA: IGI Global Publishing.

National Research Council (NRC) (1996). *National Science Education Standards*. Washington, DC: National Academy Press.

National Science Board, (1983). *Educating American for the 21st century*. Washington, D.C.: U.S. Government Printing Office.

National Science Teachers Association (NSTA). (1982). *Science-technology-society: Science education for the 1980s (NSTA Position Statement)*. Washington, DC: National Science Teachers Association.

National Science Teachers Association (NSTA), (1990-91). The NSTA position statement on science-technology-society (STS). In *NSTA handbook* (pp. 47-48). Arlington, VA: NSTA Press.

Nbina, J.B. & Obomanu, B.J. (2010). The Meaning of Scientific Literacy: A Model of Relevance in Science Education. *Academic Leadership*: *The Online Journal*, *8*(4). Retrieved May 3, 2011 from *http://www.academicleadership.org/pdf/ALJ_ISSN1533-7812_8_4_863.pdf.*

Nelson, S. A. (2010). *Weathering and clay minerals*. Retrieved June 28, 2011 from *http://www.tulane.edu/~sanelson/eens211/weathering&clayminerals.htm*

New London Group. (1996). A pedagogy of multiliteracies: Designing social futures. *Harvard Educational Review*, *66*(1), 60-92.

Ng, W. (2006). Web-based technologies, technology literacy and learning in Tan, L.W.H. and Subramaniam, R. (Eds.), *Handbook of Research on Literacy in Technology* at the K1-2 Level (pp. 94-117). Hershey, PA: Idea Group Publishers.

Ng, W. (2008). Self-directed learning with web-based sites: How well do students' perceptions and thinking match with their teachers? *Teaching Science*, *54*(2), 26-30.

Ng, W. (2009). Nanoscience and nanotechnology for the middle years. *Teaching Science*, 55(2), 16- 24.

Ng, W. (2010). Empowering students to be scientifically literate through digital literacy in Rodrigues, S. (Ed.), *Multiple Literacy and Science Education: ICTs in Formal and Informal Learning Environments* (pp. 11-31). Hershey, USA: IGI Global Publishing.

Ng, W. and Gunstone, R. (2002) Students' perceptions of the effectiveness of the World Wide Web as a research and teaching tool in science learning. *Research in Science Education*, *32*(4), 489-510.

Ng, W. & Gunstone, R. (2003) Science and Computer-based Technologies in Victorian Government Schools: Attitudes of Secondary Science Teachers.

Journal of Research in Science and Technology Education, *21*(2), 243-264.

Ng, W. & Nicholas, H. (2007). Ubiquitous learning with handhelds in schools. In Norman, A. & Pearce, J. (Eds.), *Proceedings of the 6th International Conference on Mobile Learning,* Melbourne, Australia, pp. 186-193.

Ng, W. & Nicholas, H. (2009). Introduction of pocket PC in schools: attitudes and beliefs in the first year. *Computers and Education*, *52*(2), 470-480.

Ng, W. & Anastopoulou, S. (2011). Formal and informal use of handhelds by Australian and British students: A comparative case studies in W. Ng (Ed.), *Mobile Technologies and Handheld Devices for Ubiquitous Learning: Research and Pedagogy* (pp. 279-298). Hershey, PA: IGI Global Publishing.

Ng, W. & Nicholas, H. (2011). Insights into students' thinking with handheld computers in W. Ng (Ed.), *Mobile technologies and handheld devices for ubiquitous learning: Research and pedagogy* (pp. 79-98). Hershey, PA: IGI Global Publishing.

Ng, W., Nicholas, H., Loke, S. & Torabi, T. (2010). Designing effective pedagogical systems for teaching and learning with mobile and ubiquitous devices in Goh, T. T. (Ed.), *Multiplatform E-Learning Systems and Technologies: Mobile Devices for Ubiquitous ICT-Based Education* (pp. 42-56). Hershey, PA: IGI Global Publishing

Nicholas, H. (2011). Ubiquitous computing does not guarantee ubiquitous learning in schools: The case of handheld computers in W. Ng (Ed.), *Mobile technologies and handheld devices for ubiquitous learning: Research and pedagogy* (pp. 30-44). Hershey, PA: IGI Global Publishing.

Nicholas, H. & Ng, W. (2009). Ubiquitous learning and handhelds: An overview of theory and pedagogy. In Rogers, P., Berg, G., Boettcher, J., Howard, C, Justice, L. & Schenk, K. (Eds.), *Encyclopedia for online and distance learning, Volume IV* (pp. 2171-2176). Hershey, PA: IGI Global Publishers.

Nixon, H., Atkinson, S. & Beavis, C. (2006). New media pathways: Navigating the links between home, school and the workplace in Tan, L.W.H. and Subramaniam, R. (Eds.), *Handbook of Research on Literacy in Technology at the K1-2 Level* (pp. 118-136). Hershey, PA: Idea Group Publishing.

Norris, S. & Phillips, L. (2003). How literacy in its fundamental sense is central to scientific literacy. *Science Education*, *87*(2), 224–240.

Novak, J. D. & Cañas, A. J. (2008). The Theory Underlying Concept Maps and How to Construct and Use Them, Technical Report IHMC

CmapTools 2006-01 Rev 01-2008, Florida: Florida Institute for Human and Machine Cognition. Retrieved July, 22, 2011 from *http://cmap.ihmc.us/Publications/ResearchPapers/TheoryUnderlyingConceptMaps.pdf*

Nuthall, G. (1999). The way students learn: Acquiring knowledge from an integrated science and social studies unit. *Elementary School Journal, 99*(4), 303–341.

Oblinger, D. G. (2003). Boomers & gen-Xers, millennials: Understanding the "new students". *EDUCAUSE Review*, *38*(4), 37–47.

OECD. (2003). *The PISA 2003 Assessment Framework – Mathematics, Reading, Science and Problem Solving Knowledge and Skills*. Paris: OECD.

Oliver, R. & Tomei, L. (2000). Information and communications technology literacy: getting serious about IT. *Proceedings of ED-MEDIA 2000. World Conference on Educational Multimedia, Hypermedia and Telecommunications, Virginia*. Retrieved May 12, 2011, from *http://elrond.scam.ecu.edu.au/oliver/2000/emict.pdf*

Paivio, A (1986). *Mental representations: a dual coding approach*. Oxford. England: Oxford University Press.

Palm Inc. (2002). *Palm Education Pioneers Program Final Evaluation Report.* Retrieved March 21, 2011 from *http://www.palmgrants.sri.com/PEP_Final_Report.pdf*

Papert, S. (1980). *Mindstorms*. New York: Basic Books.

Papert, S. (1993). *The children's machine: Rethinking school in the age of the computer*. New York: Basic Books.

Parliament of Victoria, Education and Training Committee. (2006). *Inquiry into the Promotion of Mathematics and Science Education: Final report.* Victoria: Victorian Government Printer.

Paul, P.V. (2006). New Literacies, Multiple Literacies, Unlimited Literacies: What Now, What Next, Where to? A Response to Blue Listerine, Parochialism and ASL Literacy. *Journal of Deaf Studies and Deaf Education*, *11*(3), 382-7.

Pedretti, E., Mayer-Smith, J. & Woodrow, J. (1998). Technology, text and talk: Students' learning in a technology enhanced secondary science classroom. *Science Education*, *82*(5), 569-589.

Pella, M. O., O'Hearn, G. T., & Gale, C. G. (1966). Referents to scientific literacy. *Journal of Research in Science Teaching*, *4*, 199–208.

Perry, D. (2003). *Handheld computers (PDAs) in schools*. Coventry: BETCA.

Piacciano, A. (2001). *Distance learning*. New Jersey: Merrill, Prentice Hall.

Piaget, J. (1955). *The construction of reality in the child.* London: Routledge & Keegan Paul.

Piaget, J. (1972). *Psychology and epistemology: Towards a theory of knowledge.* London: Penguin University Books.

Pittard, V., Bannister, P. & Dunn, J. (2003). *The Big pICTure: The Impact of ICT onAttainment, Motivation and Learning.* UK: DfES Publications

Prain, V., & Waldrip, B. (2006). An exploratory study of teachers' and students' use of multimodal representation of concepts in primary science. *International Journal of Science Education, 28*(15), 1843-1866.

Prain, V., & Waldrip, B. (2010). Representing science literacies: An introduction. *Research in Science Education, 40*(1), 1-3.

Prensky, M. (2001). Digital Natives, Digital Immigrants. *On the Horizon, 9*(5). Retrieved May 12, 2011 from *http://www.marcprensky.com/writing/prensky%20%20digital%20natives, %20digital%20immigrants%20-%20part1.pdf.*

Punie, Y., Zinnbauer, D. & Cabrera, M. (2006). *A review of the impact of ICT on Learning.* Luxembourg: Office for Official Publications of the European Communities

Raizen, S., Sellwood, P., Todd, R., & Vickers, M. (1995). *Technology education in the classroom.* San Francisco, CA: Jossey-Bass Publishers.

Raja, D., Bowman, D., Lucas, J., & North, C. (2004). Exploring the Benefits of Immersion in Abstract Information Visualization. *Proceedings of the 8th International Immersive Projection Technology Workshop, Iowa State, IA.*

Ramey-Gassert, Linda, Walberg III, H.J., & Walberg, H.J. (1994). Reexamining connections: Museums as science learning environments. *Science Education, 78*(4), 345-363

Read, T. (2000). Kormilda Science Project – Part 2. *Labtalk, 44*(3), 22-27.

Reader, J. (1988). *Missing Links.* London: Penguin Books

Reimann, P. and Goodyear, P. (2004). *ICT and Pedagogy Stimulus Paper, version 3.3.* Retrieved April 15, 2011 from *http://lrnlab.edfac.usyd.edu.au:8200/Research/mceetya2004/report/Archive/ExFile.2004-06-15.8659319410/attach/rep33.doc*

Reinking, D. (1998). Synthesizing technological transformations of literacy in a post-typographicworld. In D. Reinking, M.C. McKenna, L.D. Labbo, & R.D. Kieffer (Eds.), *Handbook of literacy and technology: Transformations in a post-typographic world* (pp. xi–xxx). Mahwah, NJ: Erlbaum.

Rennie, L. J. & Williams, G.F. (2002). Science Centers and Scientific Literacy: Promoting a Relationship with science. *Science Education, 86*(5), 706-26.

Roblyer, M.D., & Doering, A.H. (2010). *Integrating educational technology into teaching*. (5th ed.). Boston: Allen and Bacon/Pearson.

Rogers, A. (1997). Learning: Can we change the discourse? *Adults learning, 8*(5), 116-117.

Rogers, A. (2005). *Non-formal education: flexible schooling or participatory education?* Hong Kong Comparative Education Research Centre: Kluwer Academic Pub.

Romeo, G.I. (2006). Engage, empower, enable: Developing a shared vision for technology in education in D. Hung & M.S. Khine (Eds.), *Engaged learning with emerging technologies* (pp 149-176). Dordrecht: Springer.

Russell, T. & McGuigan, L. (2001). Promoting understanding through representational redescription: an illustration referring to young pupils' ideas about gravity. In D. Psillos, P. Kariotoglou, V. Tselfes, G. Bisdikian, G. Fassoulopoulos, E. Hatzikraniotis, & E. Kallery (Eds.), *Science education research in the knowledge-based society. Proceedings of the Third International Conference of the ESERA* (pp. 600–602). Thessaloniki, Greece: Aristotle University of Thessaloniki.

Sagan, C. (1997).*The demon-haunted world: science as a candle in the dark*. New York: Ballantine Books.

Sargant, N. (1991). *Learning and 'Leisure': A study of adult participation in learning and its policy implica tions*. Leicester: National Institute of Adult Continuing Education.

Schreiner, C. & Sjøberg, S. (2004). *Sowing the seeds of ROSE: Background, Rationale, Questionnaire Development and Data Collection for ROSE (The Relevance of Science Education) - a comparative study of students' views of science and science education*. Dept. of Teacher Education and School Development, University of Oslo, Oslo. Retrieved July 20, 2011 from *http://www.ils.uio.no/english/rose/publications/english-pub.html*

Shamos, M. H. (1995). *The myth of scientific literacy*. New Brunswick, NJ: Rutgers University Press.

Shepard, S. (2004). *Managing the millennials*. Retrieved July 24, 2010, from *http://www.shepardcomm.com/managing-millennials-wp.pdf*

Shen, S.P. (1975). Science literacy and the public understanding of science. In S.B. Day (Ed.), *Communication of Scientific Information* (pp 44-52). Karger: Basel.

Shih, J.L., Chuang, C. W. & Hwang, G. J. (2011). Historic monument education: The impact of a collaborative inquiry-based mobile learning strategy on social relationship development. In W. Ng (Ed.), *Mobile technologies and handheld devices for ubiquitous learning: Research and pedagogy* (pp. 128-169). Hershey, PA: IGI Global Publishing.

Shulman, L.S. (1987). Knowledge and teaching: foundations of the new reform. *Harvard Educational Review*, *57*(1), 1-22.

Siemens, G. (2005). Connectivism: a learning theory for the digital age. *International journal of instructional technology & distance learning*, *2*(1), 3-9.

Sinitsa, K. (2000). Learning individually: a life-long perspective. *Educational Technology & Society* , *3*(1), 17-23.

Smith, E. (2010). Is there a crisis in school science education in the UK? *Educational Review*, *62*(2), 189-202.

Smith, M. (1996). Self-direction. *The encyclopaedia of informal education*. Retrieved June 28, 2011 from *http://www.infed.org/biblio/b-selfdr.htm.*

Smith, P., Mahdavi, J. Carvalho, M. & Tippet, N. (2006). *An investigation into cyber bullying, its forms, awareness and impact, and the relationship between age and gender in cyber bullying: A report to the anti-bullying alliance.* London. UK: DfE. Retrieved May 12, 2011, from *https://www.education.gov.uk/publications/eOrderingDownload/RBX03-06.pdf*

Sprague, D., Kopfman, K., & Dorsey, S. (1998). Faculty development in the integration of technology in teacher education courses. *Journal of Computing in Teacher Education, 2*(14), 24-28.

Stewart, K.M. & Hedberg, J.G. (2011). The pedagogy of mobility. In W. Ng (Ed.), *Mobile technologies and handheld devices for ubiquitous learning: Research and pedagogy*, (pp.258-278). Hershey, PA: IGI Global Publishing.

Stocklmayer, S. M. , Rennie, L. J. & Gilbert, J. K.(2010). The roles of the formal and informal sectors in the provision of effective science education. *Studies in Science Education, 46*(1), 1-44.

Strategic ICT Advisory Service (2009). *Hot topic: ICT in preservice teacher training*. Retrieved July 2, 2011 from *http://dspace.edna.edu.au/dspace/bitstream/2150/54714/1/SICTAS_HT_pre-service.pdf*

Stylianou D. & Pitta-Pantazi D. (2002). Visualization and high achievement in mathematics: A critical look at successful visualization strategies. In F.

Hitt (Ed.), *Representations and mathematics visualization* (pp. 31- 46). Mexico: Cinvestav-IPN.

Sweller, J.(1988). Cognitive load during problem solving: Effects on learning. *Cognitive Science*, *12*(2), 257-285.

Sweller, J.(2005). Implications of cognitive load theory for multimedia learning. In R.E. Mayer (Ed.), *The Cambridge Handbook of Multimedia Learning* (pp. 19-30). New York: Cambridge University Press.

Sweller, J. (2009). Cognitive bases of human creativity. *Educational Psychology Review*, *21*(1), 11-19.

Sweller, J., van Merrienboer, J.J.G. and Paas, F.G.W.C. (1998). Cognitive architecture and instructional design. *Educational Psychology Review*, *10*(3), 251-295.

Tasmanian Department of Education (2009). *Critical Literacy*. Retrieved May 12, 2011 from *http://www.education.tas.gov.au/curriculum/standards/english/english/teachers/critlit*

Tebbutt, M. (2000). ICT in science: problems, possibilities… and principles? *School Science Review*, *81*(297), 57-64.

The European White Paper on Education and Training, (1995). Retrieved April 12, 2011 from *http://europa.eu/documents/comm/white_papers/pdf/com95_590_en.pdf*

The Royal Society, (1985). *The public understanding of science*. London: The Royal Society.

Tough, A. M. (1967). *Learning without a teacher. A study of tasks and assistance during adult self-teaching projects*. Toronto: Ontario Institute for Studies in Education.

Town, S. (2000). Wisdom or welfare? The seven pillars model. In S. Corrall & H. Hathaway (Eds.), *Seven pillars of wisdom? Good practice in information skills development* (pp. 11-21). London: SCONUL.

Treder, M. (2004). *Jolt to the system: the transformative impact of nanotechnology*. Retrieved July 12, 2011 from *http://www.cmano.org/Speech%20%20Troy%20%20Mike%20Treder,%20CRN.ppt*

Tuman, S. M. (2008). Guiding the design of E-learning programs: managing presentation content to match the cognitive factors of users. In T. B. Scott & J. I. Livingston (Eds.), *Leading edge educational technology* (pp. 49-75). New York: Nova Science Publishers.

Tyner, K. (1998). *Literacy in a digital world: Teaching and learning in the age of information*. Mahwah, NJ: Lawrence Erlbaum Associates.

Tytler, R. (2007). *Re-imagining science education: Engaging students in science for Australia's future.* Camberwell, Victoria: ACER Press.

University of Cambridge (2010). *Mobile communications for medicine.* Retrieved July 12, 2011 from *http://www.cam.ac.uk/research/news/mobile-communications-for-medicine/*

Upper Yarra Community House (not dated). *Integrating new literacies into classroom practise final report.* Retrieved May 12, 2011 from http://www.newliteracies.com.au/file/file/Integrating%20New%20Lits%20Into%20Classroom%20Practice%20Final%20Report.pdf.

van 't Hooft, M. (2006). Tapping into digital literacy: Handheld computers in the K-12 classroom. In L.W.H. Tan & R. Subramaniam (Eds.), *Handbook of Research on Literacy in Technology at the K-12 Level* (pp. 287- 307). Hershey, USA: Idea Group Publishing.

Vygotsky, L. (1978). *Mind in society: The development of higher psychological processes*. Cambridge, MA: Harvard University Press.

Vygotsky, L. S. (1962). *Thought and language*. Cambridge, MA: MIT Press.

Waldrop, M. M. (2008). Science 2.0 -- Is Open Access Science the Future? *Scientific American*. Retrieved July 20, 2011 from *http://www.scientificamerican.com/article.cfm?id=science-2-point-0*

Waldrip, B., Prain, V. & Carolan, J. (2006). Learning junior secondary science through multi-modal representations. *Electronic Journal of Science Education*, *11*(1), 87-107.

Wallace, R. (2002). The internet as a site for changing practice: the case of Ms Owens. *Research in Science Education*, *32*(4), 465-487.

Wallace, R., Kupperman, J., Krajcik, J. & Soloway, E. (2000). Science on the Web: Students on-line in a sixth grade classroom. *Journal of Learning Sciences*, *9*(1), 75-104.

Waycott, J., Jones, A., & Scanlon, E. (2005). PDAs as lifelong learning tools: An activity theory based analysis. *Learning, Media and Technology*, *30*(2), 107-130.

Wayne, M. & Cooper, L. (2007). A white paper on Digital inclusion imperatives offer municipalities new social and economic opportunities. Retrieved June 9, 2011 from *http://www.digitalaccess.org/pdf/White_Paper.pdf*

Webb, M. (2005). Affordances of ICT in science learning: implications for an integrated pedagogy. *International Journal of Science Education*, *27*(6), 705–735.

Weiser, M. (1991). The Computer for the Twenty-First Century, *Scientific American*, September issue, 94-10.

Weiser, M., & Brown, J. S. (1996). *The coming of age of calm technology.* Retrieved August 2, 2011 from *http://www.ubiq.com/hypertext/weiser/acmfuture2endnote.htm*

Willis, J. & Cifuentes, L (2005). Training teachers to integrate technology into the classroom curriculum: Online versus face-to-face course delivery. *Journal of Technology and Teacher Education, 13*(1), 43-64.

Wolak, J., Mitchell, K., & Finkelhor, D. (2006). *Online victimization of youth: five years later.* National Center for Missing & Exploited Children bulletin - #07-06-025. Alexandria, VA.

Wynne, M. E. & Cooper, W. L. (2007). *Power up: The campaign for digital inclusion.* Retrieved May 12, 2011 from *http://www.digitalaccess.org/pdf/White_Paper.pdf*

Yoder, M.B. (1999). A productive and thought-provoking use of the Internet. *Learning and Leading with Technology*, *26*(7), 6-11.

Yu, K.C. (2005). Digital Full-Domes: The Future of Virtual Astronomy Education. Planetarian, *Journal of the International Planetarium Society, 34*(3), 6 -11.

Zurita, G. and Nussbaum, M. (2004). Computer supported collaborative learning using wirelessly interconnected handheld computers. *Computers & Education*, *42*(3), 289-314.

INDEX

C

G

H

I

J

K

L

M

N

R

S